Praise for *Sun, Sea, Soi*

T0279429

"For the twenty years I've written about East Coast wines, no one has taught me more about them than Rich Olsen-Harbich has. It's not surprising that in distilling his forty-plus years of grape growing and winemaking experience into *Sun, Sea, Soil*, he has delivered the most important and comprehensive book ever written about Long Island wine. In it, he examines the district literally from the ground up, exploring its history, meteorology, hydrology, pedology, viticulture, and enology—but really, this book is a love letter to the wine region and community where he's made it his life's work to define the unique style of Long Island wine."

— Lenn Thompson, owner, Cork Report Media

"I had the good fortune of entering the New York wine industry at the same time as Rich Olsen-Harbich, and of witnessing the winegrowing revolution on Long Island, thanks largely to his passion, leadership, and collaboration. To the book's title I would only suggest adding the word 'Soul' because so much of this inspiring tale chronicles the vital role of people—women, Latinos, true believers in Long Island's future—in transforming a landscape of potato fields into a world-class wine region. Cheers!"

— Jim Trezise, president of WineAmerica and former president of the New York Wine and Grape Foundation

"An extraordinary tour de force that should be required reading for anyone interested in wine—and its humanity. Simply the most rigorously conceived book ever written about the modern American wine industry, by one of its most esteemed winemakers and intellectuals. Richard Olsen-Harbich has crafted an elegant meditation on the complexities of wine and its relationship to American culture. He tells rich stories of place while ignoring the borders dividing genres, deftly weaving together vibrant personal experiences with revelatory facts of science and history. His efforts to situate the beauty, oddity, and hedonism that distinguish wine from all other forms of agriculture are both critical and compelling."

—Trent Preszler, CEO of Bedell Cellars and Professor of Practice at Cornell University

Sun, Sea, Soil, Wine

Sun, Sea, Soil, Wine

Winemaking on the North Fork of Long Island

RICHARD OLSEN-HARBICH

EXCELSIOR
EDITIONS

Cover: Background photo courtesy Bedell Cellars.

Published by State University of New York Press, Albany

Excelsior Editions is an imprint of State University of New York Press

For information, contact State University of New York Press, Albany, NY
www.sunypress.edu

Library of Congress Cataloging-in-Publication Data

Name: Olsen-Harbich, Richard, author.
Title: Sun, sea, soil, wine : winemaking on the North Fork of Long Island /
    Richard Olsen-Harbich.
Description: Albany, NY : State University of New York Press, [2024] | Series: Excelsior
    editions | Includes bibliographical references and index.
Identifiers: LCCN 2023009045 | ISBN 9781438495521 (pbk. : alk. paper) | ISBN
    9781438495514 (ebook)
Subjects: LCSH: Wine and wine making—New York (State)—North Fork
    (Peninsula) | Vineyards—New York (State)—North Fork (Peninsula)
Classification: LCC TP557.5.N7 O47 2024 | DDC 663/.200974725—dc23/eng/20230302
LC record available at https://lccn.loc.gov/2023009045

10 9 8 7 6 5 4 3 2 1

In loving memory of my Father and Mother,
Alfred and Lottie Harbich,
who always provided unconditional love and support.
In the process, they raised a winemaker.

༒

For Nancy, Emily, and Peter.
My family, my heart, and my terroir.

I was born within the sound of the sea—down on Long Island and I know all the songs that the seashell sings.

—Walt Whitman

Long Island Sound. Courtesy of Denise Winter-Peragallo.

Contents

Part 2. Spontaneous Fermentation

Acknowledgments

For me, writing a book is more complicated than making wine. The work presented took years to complete, and I'm sure I still left a lot out, but the process was more rewarding than I ever imagined. Unlike writing, winemaking is a team sport; so many people have been involved over the years in helping me create beautiful wines.

First and foremost, none of this would have been possible without my wife and best friend, Nancy, who has accompanied me on my winemaking journey every step of the way. She's my sounding board for new ideas and taste tester for every one of my creations. Without her constant love and support, I don't think I'd be here today.

Several other people deserve my heartfelt thanks:

Lyle Greenfield, who took a chance on a twenty-two-year-old kid and let me run his vineyard and winery in Bridgehampton. He saw someone who wanted to learn, grow, and succeed in the wine business. He always believed in me and pushed me to be the best I could be. He also helped me with the title of this book.

Trent Preszler, CEO of Bedell Cellars and our owner Ninah Lynne, who encourage my winemaking journey at Bedell and continue to believe in me, providing me with the support and creative freedom to make great wines and explore the complexities of our terroir. I can't imagine a better place to make wine.

Marin Brennan, my greatest student and protégé to whom I leave my legacy of winemaking, and my loyal and dedicated cellar staff, who always lift me up and continue to make me proud. Without their assistance during this past year, I wouldn't have been able to complete this manuscript.

Carlo DeVito, an accomplished author, and winemaker in his own right, who put in the good word for me at SUNY Press and sent me on my writing journey. I don't think this book would have happened without his support.

Stephen Mudd, for giving me my first job in a vineyard and teaching me what vines need to succeed on the North Fork, making the most out of every vintage. I not only learned how to grow wine, but I also gained a brother.

Larry Perrine and Alice Wise, for pushing growers and winemakers to be the best they could be through thoughtful research and investigation, helping to make Long Island the world-class wine region it is today. In the process of adding to my intellectual development, they became my lifelong friends.

Many thanks to everyone on the SUNY Press team who helped me make this book a reality, especially Richard Carlin, Susan Geraghty, and Céline Parent for their expert guidance.

Finally, to all those who have played a large part in my journey and are no longer with us: Michael Lynne, Jonathan Lynne, Don Cavaluzzi, David Mudd, Jack Petrocelli, Paul Pontallier, Ben Sisson, and Donna Rudolph.

My greatest wish in writing this book is to help people understand the unique beauty of the North Fork and the wines that are made there.

Mattituck, NY, 2022

Eastern Long Island. Courtesy of Google Maps.

Prologue

The meaning of life is to find your gift. The purpose of life is to give it away.

—Pablo Picasso

It was the light. The way it reflected over the open fields and seemed to bounce over the ground. That was the main thing I noticed about my first visit to the North Fork. I'd been to many places, but the North Fork was different. I felt the magic, smelling the sea air and the freshly tilled soil blown by a gentle breeze that filled my soul.

The North Fork has been an important farming area for centuries, first for Native Americans and, starting in the 1600s, for the early colonists. There is a solace to the place and a long history going back to the founding days of our nation. Some of the oldest houses and farms in the country still survive there, whispering the stories of people from another time. Many families have worked the land for generations, and it remains a vital farming community to this day, generating some of the highest agricultural revenues in New York. The conditions are idyllic for growing crops with deep, fertile soils with few rocks to impede cultivation. On some farms, the land rolls right down to the shoreline.

The land from Wading River to Orient Point is blessed with a gentle, maritime climate that is neither too cold in winter nor too hot in summer. It is genuinely a Goldilocks place—a bucolic oasis in an otherwise harsh northern zone less than one hundred miles east of the largest metropolis in North America. It's a miracle it exists at all.

I've always loved the outdoors. As a child growing up in New Hyde Park, within walking distance of the Queens border, the thought of living in open spaces always felt liberating. I wasn't raised in a wine-centric home, although wine flows through the veins of my family history.

My paternal grandfather came to New York City from Austria in 1912. He played the piano and found work in silent movie theaters while also laboring as a mechanic in a shipyard during the First World War. Once the war was over, and the "talkies" arrived, he worked for a tavern on Carlisle Street in Lower Manhattan. With the onset of prohibition, he helped turn the tavern into a speakeasy, which he managed for over a decade. After returning to Austria in the early thirties, he took over his family's hotel, where my father would help him source wine from nearby Moravian vineyards. Some of my father's earliest memories were of filling bottles of wine from a barrel in the tavern's cellar. He told me he used to sneak tastes when his father wasn't looking. It was his favorite job.

My mother's side of the family came from the German wine region of the Nahe. She was born in a house surrounded by hillside vineyards, and her parents worked for local winemakers while they were growing up. As I like to say, wine runs through my family's genes. It didn't take too much to turn on the switch.

My earliest recollection of wine was from a homemade batch my maternal grandfather made from red currants. He grew a small row of these bushes on the side of our garage, and every summer, he would pick and crush the berries into a small glass carboy. For a few weeks, the clear jug of pink liquid would ferment in our basement, and I would sit and watch the water bubble and burp in the glass fermentation lock. In time, the bubbles would become less frequent, and eventually, my grandfather would take a rubber hose and pour some into a small glass. I can recall the taste—fizzy and sour—and I remember hearing my grandfather laugh as I ran back upstairs to get something to wash it down. Growing up, the best wines I remember weren't from the local store; they were brought home from Germany after my parents or grandparents visited relatives. I had to wait until a Thanksgiving dinner or Christmas goose to taste those. Nahe wines remain my "soul food" to this day.

As far back as I can remember, I knew I never wanted to wear a suit and tie and work in an office. As a boy, I constantly sought the outdoors, whether it was in a local park or by the sea. As I got older, something inside of me pulled me outside to work with my hands and make things grow. I spent summers working at a local nursery and public garden, getting soil under my fingernails, learning as much as possible about plants, driving tractors, and being part of a team. My late mom never got over the fact that in my high school yearbook, I put the word "farmer" down under the section that listed my future career goals. It's what I wanted to be then, and it turns out it wasn't far from the truth. Making wine is like being a farmer; we're taking the crop to its next incarnation, at the mercy of the weather, climate, and countless issues beyond our control.

This desire followed me into college and, eventually, to Cornell University, where I discovered my future career and my future wife. I laid eyes on Nancy the first day we both arrived at Cornell, and I've been in love ever since. It took me a few more months to eventually fall in love with wine. I'd dabbled in agronomy and dairy science and even considered a career in tropical agriculture for a time. Still, it wasn't until I started learning about grapes that something took root in me. I remember sharing a bottle of a 1979 Hermann Wiemer Riesling with Nancy in my apartment in Ithaca, NY. It was a revelation, and I soon knew what I wanted to do. I cobbled together all the classes I could find about viticulture and enology and, along with lots of independent study, started my viticultural journey.

I first worked in vineyards during the summer of 1982 for David and Stephen Mudd, two founders of the North Fork region. My days were long and filled with hoeing weeds, tying up vines, sawing dead grapevine trunks, and anything else that needed to get done. I spent days in the hot sun riding a two-seated planter, placing new vines in the ground, one at a time, while inhaling exhaust from a tractor. Once we finished the planting, I hauled thousands of wooden posts off trucks, laid them out in the field, and one by one, set them up as straight as I could before a post-pounder hammered them into the ground. Vineyard work was strenuous and dirty, but it was incredibly gratifying.

My journey continued straight after graduation from Cornell at a small start-up in Bridgehampton on the South Fork. Here, I literally learned from the ground up, managing both the vineyard and cellar. When I arrived in early June of 1983, I found most of the twenty-five-acre vineyard sprawling over the ground and partially flooded. I had no winery equipment save for a few un-jacketed stainless-steel tanks, no tractor, no sprayer, and no crew. I couldn't have been more thrilled. The owner, Lyle Greenfield, wisely hired the highly respected Finger Lakes winemaker Hermann Wiemer as a consultant, and I worked alongside Hermann for two years, honing my craft. I made just two stainless steel white wines for the first few years—a Chardonnay and a Riesling. It was one of the first domestic Chardonnays made this way and probably ahead of its time. Back then, many people thought we were out of our minds trying to sell a bottle of wine from Long Island for twelve dollars, but we did, and it sold out quickly. Lyle and I worked tirelessly to make the business work and had many successful years, including having one wine make the Wine Spectator Top 100. But ultimately, despite our optimism about the Atlantic's tempering breezes, the low-lying site posed insurmountable challenges for grape growing. And it didn't help that the '80s had been the coldest decade on record in the Hamptons.

Ironically, my love of working outside eventually led me away from the vineyard and into the wine cellar. I soon learned that making great wine in this part of the world takes tremendous expertise in viticulture and enology—and I was determined to succeed. Since my first vintage as a winemaker in the fall of 1983, I've worked a harvest every year—picking grapes, pressing, fermenting, and nursing them into wine through all kinds of weather and conditions.

Once the Bridgehampton Winery sold, I went to work for Alex and Louisa Hargrave on their eponymous estate. Working in the historic vineyard and cellar where our modern wine district began was thrilling and humbling and my winemaking education grew because of that experience. Over the years, I consulted for many producers, including Gristina, Palmer, Jamesport, Paumanok, Wölffer, Peconic Bay, Schnieder, Broadfields, Channing Daughters, Croteaux, and Raphael, along with several upstate producers. I was eventually hired to direct the winemaking program at Raphael in 1997. I had a prolific, fourteen-year run with that estate before being hired in 2010 to lead the winemaking effort at the iconic Bedell Cellars, where I've settled in ever since.

During the early days, there was a huge learning curve as no information on Long Island winemaking was available. The reference books I had were either from the West Coast or in another language—and there wasn't any Google translation available. I was lucky to have a great mentor in Hermann Wiemer and learned from many others along the way, including Stephen Mudd, Lyle Greenfield, Alex and Louisa Hargrave, Larry Perrine, Dan Kleck, Alice Wise, and the great Paul Pontallier of Château Margaux. Even so, it took a lot of trial and error to figure things out. As I like to tell people, no one has made more mistakes in the wine cellar than I have. Every place I've worked, I learned something new—good and bad. I learned to build on my knowledge by holding on to the techniques that worked and discarding the rest. I think it's a good lesson for life in general.

Winemakers have one chance a year to get it right; if you're unhappy with the results, you must wait an entire year to try again. Luckily, I've been given many chances to get it right and have spent my career working to define a definitive style of wine for the North Fork.

In a world where instant gratification is often not good enough, and data can travel in a millisecond, the craft of growing grapes and making wine seems downright antiquated—and it is. Winemaking is an ancient profession providing a link to the past, a panacea for the present, and hope for the future.

As I enter the autumn of my life, I felt the time was right to pass along what I've learned from one of the country's most surprising yet rewarding wine regions. I'm aware that not everyone will agree with it all, but to paraphrase The Band, you can take what you want and leave the rest. This book comprises what I've learned and come to know in over forty years of growing grapes and making wine on Long Island. This is my own experience, my awareness. And now it's yours for the taking.

Developing a new wine region and a new wine style for our country is not a task for the faint of heart. Many "experts" didn't think we would succeed here, and some naysayers still believe that. We've had to fight for respect and recognition every step of the way and continue to do so. However, a track record of success precedes this book and these words. A level of excellence has been achieved through the efforts of many dedicated and passionate people who put roots in the ground on the North Fork to create unique and world-class wines.

This fact was never more apparent than in January of 2013 when the 2009 Bedell Merlot was chosen to be served at President Barack Obama's second inaugural luncheon held at the National Statuary Hall in the Capitol Building following President Obama's swearing-in. About two hundred guests attended the luncheon, including President Obama, Vice President Joe Biden, their families, as well as cabinet members, leaders of Congress, and Supreme Court justices. It was the first time a New York wine was chosen for an inaugural event, proving that North Fork wine had truly arrived. I was enormously proud.

Although I happily worked in Bridgehampton for almost ten years and wrote the Hamptons American Viticultural Area (AVA), this book is focused primarily on the North Fork. Previous books on the Long Island wine region covered the entire East End and the few vineyards that grow wine grapes west of Riverhead. But the heart, soul, and foundation of the Long Island wine district lie within the North Fork AVA. It is where I spent most of my winemaking career and where over ninety percent of Long Island wine grapes are grown. Without the North Fork, it would be hard for the Hamptons and greater Long Island AVAs to survive and have enough critical mass to impact the wine world. As of this writing, the North

Fork of Long Island remains the leading producer of European wine grapes on the East Coast in terms of acreage and tonnage.

Making wine—great wine—requires an understanding of where you're making it. It's been my life's work to learn about the North Fork and why it works so well for wine, and this book consists of what I have discovered. It's exhilarating making wine in a brand-new region, and I'm still amazed by what our soil and climate can produce. I feel blessed to be part of this profession and believe my ancestors would be as happy as I am with my career choice. The combination of hospitality and agriculture fits right in with my genetic history. Of course, running a speakeasy isn't the same as making wine, but we were both after the same things—giving people what they want and trying to make them happy.

I originally started to write this book many years ago but gave up, thinking I didn't have enough to say. Maybe I didn't back then, but now, I hope my words will add to the historical record of the Long Island wine industry, of which I am forever a part. And maybe, like any vintage, I won't know until it's in the bottle—or between the covers, in the case of this book.

I feel it's what I need to do right now. Hopefully, it sheds light on a profession and a business that is often a mystery to people. Or maybe it'll just help my children and grandchildren get to know me better. Mainly, I hope this book will help other North Fork winegrowers along their journey. After all, if one wants to make terroir wines on the North Fork, one needs to know what our terroir is and understand its legacy in the history of winemaking from this place that I know and love. A place that is neither too hot nor too cold. A place that feels just right.

Introduction

If you want something you have never had, you must be willing to do something you have never done.

—Thomas Jefferson

In his seminal text entitled *Facing East from Indian Country: A Native History of Early America*, Dr. Daniel Richter presents an intriguing narrative of early colonial history from the Native-American point of view. The reader is asked to imagine what Native Americans thought when they first saw European settlers coming ashore as Dr. Richter turned the table on the Eurocentric version of American history. From their perspective, the native peoples lived in the familiar Old World, and the strange, white-skinned visitors were the ones from a brand-New World, building on historian Carl Becker's famous assertion that history is often "an imaginative creation."[1]

The dichotomy between the Old and New World has been ingrained in the wine lexicon for decades. The term *Old World* describes a miscellaneous category, with grape varieties, viticultural techniques, and winemaking practices adapted around their unique climates and landscapes. Old World wines are the product of European experience and taste, grown in less-than-perfect conditions dictated by agricultural evolution, terroir, and tradition.

New World wines are typically associated with grapes grown in warmer climates outside the boundaries of greater Europe, with dark, inky extraction, high levels of alcohol, and overly ripe fruit. The traditional view of New World winemaking philosophy generally places less emphasis on terroir and more on the preservation of varietal fruit character, believing that the appropriate use of science and technology in the vineyard and winery can fix any flaws and push the envelope of quality. Compared to New World wines, Old World wines tend to be lower in alcohol and hence more elegant and refreshing, with less extraction and a natural balance. Old World wines have been around for centuries and are considered New World wines' archetypes.

Like the Eastern Native Americans in Dr. Richter's text, East Coast winemakers are beginning to turn these long-held assumptions upside down. The wine districts in the region of the New World where our country began can no longer be considered "New" in the traditional sense.

American consumers have become tired of high alcohol, overly extracted wines. They want variety and choice, but mainly sincerity, purity, and something they can drink with a meal. I describe this trend as a maturation of the wine consumer; they've become secure enough in their taste buds that they no longer feel the need to impress anyone. We've witnessed an evolution of the consumer palate, leading to a more serious level of wine and food appreciation. This trend is especially robust in younger wine drinkers who are unimpressed by the traditional 100-point wine ratings from journals that are supported by large corporate advertising revenue and instead, rely on their own philosophy and palates.

The North Fork of Long Island has produced aromatic, elegant, and low alcohol wines for almost fifty years, with ever-increasing quality and little need for winemaker intervention. Our style has never wavered—we have a dedicated, monogamous relationship with crisp, aromatic whites and edgy yet elegant reds—all with moderate amounts of alcohol and a refreshing minerality. Thankfully, we don't need nor want to change to become fashionable. It's what our vineyards genuinely produce and the wine we sincerely create. It's what our terroir does all by itself. We may not fit the New World stereotype, but we'll always stay true to our identity. We have a long-term commitment to our terroir.

New York has always been where the best information, inventions, food, and fashions have come to be tested and accepted. The growing popularity of low-alcohol, cool-climate wine continues to rise in this country and is not going away. We know from our great reception in the marketplace that we have passed the test. Now we need more people to learn about what we do.

Perhaps there is a lesson we can learn from the Native Americans of Dr. Richter's text—that the New World is not quite what it seems to be. We make wine in the oldest part of America—our country's historic birth-place—in a refreshing style that is all our own. This requires a new definition along with a new *nom de plume*—one that breaks with the standard dogma and accurately describes our part of the world. It's an edgy, temperate zone with four seasons, unpredictable rainfall, humid summers, and cool ripening conditions. It's near the sea with fertile soils, mild temperatures, and lots of sunshine—all wrapped up in a distinctive New York state of wine. It's a place like nowhere else on earth.

I call it something else entirely: the Cool New World.

PART I

THE REGION

Corey Creek Vineyard. Courtesy of Steve Carlson, Bedell Cellars.

History

I too have bubbled up, floated the measureless float, and been wash'd
   on your shores,
I too am but a trail of drift and debris,
I too leave little wrecks upon you, you fish-shaped island.

—Walt Whitman

The start of modern wine growing on the North Fork began in 1973 with Alex and Louisa Hargrave. Yet the story of wine production on Long Island is older than most people think.

Wine seems to have been on the minds of the first explorers to the New World, alluding to the importance of the possibility of exports in future investments. Probably the first person to realize the potential for producing wine in this part of the world was the famous Italian navigator, Giovanni da Verrazzano. In 1524, a group of French merchants and Italian bankers hired Verrazzano, with the blessing of King Francis I of France, to find a navigable passage through America. He recorded his explorations, and his journal describes an abundance of grapevines growing into treetops, with an alluring potential for wine production. He even gave the island its first name—*Flora*.[1]

The next person to record a description of Long Island was the famed English explorer Henry Hudson who sailed for the Dutch East India Company. Departing Amsterdam in 1609 on the Dutch ship *Half Moon*, he was ordered to explore the Arctic Ocean north of Russia to find an easterly trade route to Asia. Once underway, he found the northern way blocked by ice. Not wanting to return home empty-handed, he turned his fully outfitted ship around and sailed west to seek a passage through North America. On September 3, 1609, Hudson navigated around the western end of Long

Island into what would eventually become New York Harbor and the river that bears his name. He described Long Island as "covered with forests, trees loaded with fruit and grapevines of many kinds."[2]

After Hudson's return to Amsterdam, Dutch merchants were less interested in growing wine and thought the area was worth exploring as a potential source of beaver pelts. Fur was a highly sought-after commodity in Europe, and traders were looking to meet the demand as animals became scarce in Eastern Europe and Russia. They offered trading patents to explorers who could produce accurate maps of the region and report back on the potential for the fur trade.

Looking to secure exclusive trading privileges from the Dutch States-General, a privateer named Adriaen Block made four trips to the lower Hudson region, the last one a complete circumnavigation of a large island. Block had to submit a detailed report to receive his patent and ultimately produced the first known map of the area, christening the land "Lange Eylandt." He didn't know at the time that he had circumnavigated the longest and the largest island in the contiguous United States.

The discussion about grapes appeared next in documents written to the Dutch West India Company in the mid-1600s. First published in 1655, Adriaen van der Donck's *A Description of New Netherland* was an early American sales pitch, suggesting to potential investors in New Netherland that Long Island should be considered a likely wine-producing region. He mentioned the Dutch settlers' early experiences with viticulture:

> It is unbelievable in what profusion grapevines grow wild in New Netherland; no region or corner of the country is without them. They grow on the level and open fields, in the natural forests under the trees, on the banks of rivers, creeks, and streams, on the hillsides, and on the foothills of mountains. Accordingly, wine growing has begun to be undertaken in earnest so that there are now several formal wine estates and wine hills. The good Lord has blessed these endeavors and made them bear fruit, so I am informed. Vineyards are the other. Seeing that the population of New Netherland is growing strongly now, an abundance of wine may be expected in a few short years. By bringing those means and human industry to bear on such willing forces of nature, it cannot be doubted that, in general, wines as good as from any quarter of Germany or France will be the result.[3]

Ultimately, the Dutch weren't convinced enough by these messages to invest further in the region. Instead, the original Puritan colonists extended their control of the East End region and expanded their settlements. They weren't interested in making wine; they wanted a place to practice their religion freely.

Like many areas of the New World, the North Fork was home to several communities of indigenous people before its settlement by the English. The primary groups included the Corchaugs, part of the Montaukett nation. While evidence of indigenous winemaking has yet to be found on Long Island, wild grapes were surely part of their diet. Recent studies from scientists at Wichita State University confirmed finding residue of tartaric and succinic acids on five-hundred-year-old pottery in Central Texas.[4] Since grapes are the only temperate fruit containing tartaric acid, and succinic acid is found in wine, it's quite possible that early indigenous winemaking took place in this region. Remnants of Corchaug pottery are rare, but further archaeological discoveries could lead to more studies that could answer this question. Historians have confirmed that the Corchaugs were the largest producers of wampum on the East Coast—the primary form of currency in the region during the early settlement days of the English and Dutch. The village of Cutchogue took its name from these people, which translates roughly in Algonquin to "principal place." Some historians have theorized that the name referred to the area's importance for making wampum.

According to Edna Yeager, who wrote about the history of the North Fork in the 1960s, the first settlers found that "grapes were just waiting for the winemaker." In the seventeenth century, writings suggest that quite a few people produced various types of alcohol, including wine, to sell to the early settlers. Many of these people were, in fact, squatters who did not own the land they were living on. Later, a French immigrant named Moses Fournier reportedly grew grapes in Cutchogue in the early eighteenth century, by some accounts, with help from local Corchaugs. There are records of other Fournier vineyards in Southold and Southampton on the South Fork. In his 1846 book *The Cultivation of American Grape Vines and Making of Wine*, Alden Spooner wrote, "in the early settlement of Long Island, a vineyard was cultivated near Southampton by Mr. Fournier. We understand very good wild grapes are now in great abundance in the woods and have been successfully used for wine."[5]

Despite this optimism, wine growing on Long Island did not start well. Vines from Europe (taxonomic name *Vitis vinifera*) inevitably failed, while wines made from wild native grapes (*Vitis labrusca*) left much to be desired.

The business of wine officially began with the first English governor of the New York colony, Richard Nicolls, in 1664.[6] Nicolls granted a colleague in Brooklyn the exclusive rights to grow and sell wine without taxation and implemented a levy of five shillings per acre annually for thirty years on any competing vintners. There is no record that Nicolls's plan ever came to fruition, but the arrangement would effectively prevent other early attempts to grow wine on Long Island. It took another hundred years before the topic of commercial wine production appeared again in the historical record with the Prince family and later Alphonse Loubat, who worked with wine grapes in Brooklyn and Queens.[7]

The Prince Nursery was, at one time, the largest nursery in New York. It was started in 1737 by Robert Prince and his son, William, who began with a small garden nursery overlooking Flushing Bay. As the business grew through four generations, it hugely impacted the development of the grape and wine industry in New York. Prince collected hundreds of European fruit trees and plants from Belgian and French Huguenot sources, and by 1750, William expanded the business into America's first successful commercial nursery. Its exclusive clientele included King William IV of England, George Washington, and Lewis and Clark, who added to the Prince's growing trove of botanical samples. A third-generation William Prince Jr. tripled the nursery's acreage in 1793 and christened it the Linnaean Botanic Garden, after the great Swedish botanist.[8]

In 1830, Prince wrote the first important book on winemaking in America entitled *A Treatise on the Vine: Embracing Its History from the Earliest Ages to the Present Day*. At the time of its publication, the Linnaean Botanic Garden was growing an impressive 450 varieties of grapes. In listing his inventory, Prince described the beginnings of American grape breeding. Local horticulturists selected native grapes that were found to be of high quality. And although the early European plantings never survived more than a few years, the first two hundred years of colonial settlement led to random pollination opportunities between imported European vines and nearby wild varieties. His catalog contained many of these "varieties of original native species, varieties obtained by admixture of native species, varieties obtained from seeds of exotic grapes, varieties obtained by admixture of foreign and native varieties."[9]

An enthusiastic and passionate advocate for New World winemaking, Prince reached out to others in the field to help back up his claim. In the early 1800s, as Long Island farmers became successful at growing peaches, William Prince quoted famed botanist Thomas Nuttall of Harvard University:

The peach and the vine being natural productions of the same region of the East, the opinion has been uniformly adopted, that a climate favorable to the one could not fail to be suitable to the other. And where, let me ask, does the former thrive to a greater degree than in many other sections of our country? From the shores of Long Island . . . the peach flourishes . . . hence we may deduce the most sure prospects of an equal success for the vine.[10]

Prince tried to explain why wine growing up to that time was unsuccessful in early America. "The first is the proper selection of those kinds of grapes which are suitable to the respective climates, and the second is the want of attention to the culture requisite for reopening the wood, which requires art and attention to produce the desired effect."[11] Of course, he was in the business of selling plants, so one could view these quotes as a form of marketing for the times.

Around the same time, a French inventor named Alphonse Loubat also recognized the potential of American soil for quality wine grapes. Loubat emigrated to rural Brooklyn in 1827 and helped New York City develop its first tramway. Being French, he also loved wine and lamented the lack of his favorite beverage in his new home. He claims to have planted over a thousand vines on Long Island, all imported from his father's nurseries in the Gironde. No one is sure if Loubat ever made wine to sell. Still, he did write a book on growing grapes entitled *The American Vine Dresser's Guide*, which mentioned the "particulars of that sickness of vines through the United States which prevents the European grapes from growing well here."[12]

While the European wine industry was already well established in the eighteenth century, the evolution and selection of grape varieties for American winemaking had not even begun. Efforts to make wine with native *labrusca* grapes led to poor results, producing unusual flavors, low alcohol, and overly-acidic concoctions. European growers with the finest skills and best intentions were unaware of the devastating diseases and insects that awaited their newly planted *vinifera* vines. After a few happy years of growth, these plantings eventually failed. It was clear that establishing an American wine industry would take some time.

All of these early failures—and the "sickness" alluded to by Loubat—were because European grapes were not native to our region and were highly susceptible to the American root aphid *Phylloxera*, which fed on grapevine roots, eventually killing them. Early vineyards were also subjected to several diseases, especially downy and powdery mildew, and black rot—all of which originated in eastern North America and were unknown to European vintners. Unfortunately, the challenges faced by early American vignerons would also be felt throughout Europe by the mid-nineteenth century.

The onset of faster sailing vessels and the highly fashionable (and unregulated) importation of American grapevine specimens for curious English gardeners in the late eighteenth and early nineteenth centuries set the stage for a catastrophic blow to the great vineyards of Europe. These vines were often packed in American soil for the voyage, which inadvertently carried *Phylloxera*. The aphids spread easily in the less harsh conditions of Southern England. In just a few years, *Phylloxera* destroyed most British *vinifera* vines, and through additional importation, the American pests eventually arrived on European soil. In less than fifty years, they devastated most of the continent's wine industry.

Only after decades of French and American research could European vineyards recover and American commercial winemaking truly begin. The solutions came from grafting European vines onto phylloxera-resistant American rootstocks and developing fungicidal materials to control mildew. Today, almost all European vines grown worldwide are grafted onto American-bred rootstocks and require some form of fungicidal control to combat American-born mildews.

Even after figuring out how to control *Phylloxera* and mildew, backyard arbors were pretty much the extent of grape growing on Long Island until the middle of the twentieth century. While native and hybrid plantings became the foundation for the nascent upstate wine industry, Long Island farmers focused on growing food crops to supply the nearby New York City market. Island farms became known for vegetables, fruit, grain, and potatoes by the late-nineteenth century.

Not much attention was given to growing grapes again until around 1963, when John Wickham, a descendant of the first North Fork settlers and a fruit grower in Cutchogue, planted a selection of vines for a Cornell University field trial, which included many *vinifera* table grapes. Wickham farmed the same land as the Corchaugs did centuries before him and maintained this planting for decades.

The success of John Wickham led some curious researchers at Cornell to take an interest in the North Fork. Famed Cornell horticulturist Dr. John Tomkins held conferences in the area in 1968 and 1971. At the urging of Dr. Konstantin Frank, who was the first vintner to plant European grapes in New York's Finger Lakes, Tompkins researched weather data and was convinced that growing *vinifera* grapes on the North Fork was a real possibility. In the *Suffolk County Agricultural News* in 1971, he wrote, "there are many good sites for grapes on Long Island. Some apple and dairy farmers are taking a careful look at grape-growing opportunities."[13]

It was also Dr. Tomkins who steered the Hargraves to Long Island. Hargrave Vineyard was planted in 1973 and was the first successful commercial *vinifera* vineyard on the North Fork. David Mudd and his son Stephen planted vines the following year. These first plantings in the mid-1970s led to over 2,500 acres of European wine grapes grown on eastern Long Island in the last fifty years, with over 2,100 acres planted on the North Fork.

It has taken centuries—from early colonial failures and backyard arbors to the creation of a thriving wine region, but this success was foreseen by many. The story of growing grapes and making wine on Long Island goes back to the beginnings of our country and involves countless players. Advancements in grafting and disease-controlling materials were pivotal to the development of the American wine industry and to the ability of wine growers worldwide to deal with the results of the New World "Columbian Exchange." The first wine growers of Long Island may not have been successful, but the idea of establishing a thriving wine region turned out to be prescient—albeit 350 years later. Today, when we sip a glass of wine from the North Fork, we appreciate our terroir and are filled with droplets that tell the story of this long and impressive local history.

The Land

Land really is the best art.

—Andy Warhol

Long Island has a fascinating geological history involving glacial movement, making the land beneath our feet younger than the rest of New York. The Island's unique natural elements make it perfect for fine winemaking. While the barrier beaches on Long Island's south shore formed through the behavior of wind and waves, the geologic origins of the longest and largest island in the contiguous United States are a lot more complicated.

The North Fork and the rest of Long Island make up the low country of New York State and southern New England known as the Outer Lands. These Outer Lands form the northeasternmost extension of North America's Atlantic coastal plain. Most of Upstate New York and New England is higher in elevation than Long Island; the Connecticut, Housatonic, and Thames Rivers all empty into the Long Island Sound. Yet unlike the Old World, our low country basin is located south of the highlands, reaping the benefits of warmer temperatures and the Gulf Stream.

The North Fork is also unique in that many wine regions around the world grow their grapes on land formerly covered by ancient inland seas, leaving rich and diverse "soil legacies" once the water receded. On the North Fork, this dynamic is reversed as our land was created by a glacier that slowly filled in the sea with earth from hundreds of miles away. Long Island is the product of dozens of different soil types, gathered and mixed together from as far away as northern Canada.

Over 400 million years ago, the creation of metamorphic bedrock shaped the foundation upon which Long Island rests. Seventy million years ago, an additional land mass formed on top of this bedrock from the deposition of sands and clays during the late Cretaceous period. For a time, this prehistoric land mass supported an environment that included dinosaurs and, later, large land mammals during the Pleistocene epoch. We know little about these creatures as the fossil record in these layers lies buried more than one thousand feet underground. Small areas are visible in western Queens, Brooklyn, and Manhattan, where the rocky bedrock supports skyscrapers. Exposed Cretaceous outcrops on eastern Long Island aren't visible, but this layer remains an essential part of Long Island's topography; it is the rock upon which we were built.[1]

In geological terms, the formation of Long Island as we see it today was relatively quick compared to other parts of the world. While the layer of bedrock underneath the Island is almost half a billion years old, the amount that we see, live, work, and farm on is geologically a mere newborn.

The formation of the Long Island that we know today began during the last ice age during the beginnings of the glacial period. The massive amount of snow and ice making up the late Wisconsin Glacier was the size of a continent and averaged around three thousand feet thick in the middle and 1,500 feet high at its southern end—taller than the Empire State Building. Few things in our world have been as powerful, and fewer have moved as slowly. Inch by inch, the ice worked like a giant plow over thousands of years as it traveled south, scraping off an average of sixty-five feet from the surface of southern Canada, New England, and the Appalachians. Large quantities of rock, ground into smaller particles, were carried along with the ice until it stopped around 55,000 years ago.[2]

Like the pile of snow and ice the town plow leaves across your driveway, the southern advance of the Wisconsin Glacier covered the Cretaceous layer and created the land that makes up modern Long Island. As it melted and retreated, the material in front of the glacier left behind a terminal moraine ridge. The first stop on the glacial tour is known today as the Ronkonkoma Moraine and extends from around Roslyn Heights in Nassau County, across central Long Island to the South Fork and Montauk Point. About 21,000 years ago, another thaw followed and the ice retreated further, stopping to create the Roanoke Point Moraine, which runs from Port Jefferson to Orient Point, marking the northern boundary of the North Fork. The ridges of both glaciers make up the highest elevation on the Island and range from around 250 feet to 401 feet at its highest point.[3]

On the North Fork, the moraine reaches about two hundred feet high near the Long Island Sound in Riverhead and decreases in elevation towards Peconic Bay. Most of the region's farmland lies along the middle of the peninsula, between forty–sixty feet above sea level.[4]

During these glacial pit stops, melting ice created enormous quantities of water that carried sand, silt, clay, and gravel down the ridges and over a broad plain in front of the retreating behemoth. These glacial outwash plains were laid over the bedrock base of Long Island. As the ice kept melting, a channel formed by flowing fresh water gradually filled up and became what we know today as Long Island Sound. Further melting and runoff deposited silt, clay, and fine sand to varying depths, making up the region's topsoil. Around 11,000 years ago, the fish-tailed sculpture of Long Island was essentially completed.[5]

Today, the 101,440 acres of land on the North Fork comprise a unique peninsula with Long Island Sound to the north, Peconic Bay to the south, and the Atlantic Ocean to the east. The western boundary is the line separating Brookhaven and Riverhead Townships. The North Fork is six miles wide at its widest point and a half-mile wide at its narrowest. The landscape is relatively flat compared to the mainland, with the northern and central ridges of the terminal moraine sloping gently south towards the Peconic Bay and the Atlantic Ocean, respectively. Unlike many crude maps showing the North Fork jutting horizontally to the east, the peninsula angles towards the northeast. This orientation is vital as weather arriving from the north and west travels over Long Island Sound, which highly moderates temperatures

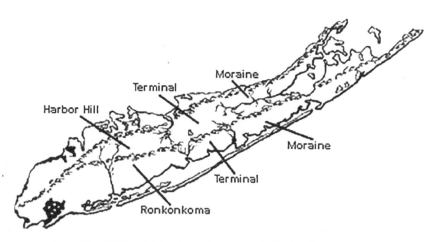

Long Island Glacial Moraines. Courtesy of Garvies Point Museum.

before reaching land.[6] This fact alone puts the North Fork in a class by itself when compared to the rest of the Eastern Seaboard.

The land making up the North Fork and the rest of Long Island lie over bedrock created during the Cretaceous period giving the Island a solid foundation. A few areas of this framework remain visible offshore and underwater. Still, the land we live and farm on is a geologically nascent, ten-thousand-year-old *pesto* of rock and gravel mixed with organic matter and laid over glacial sediment. This rich, fertile "skin" known as our soil carries the lifeblood of our region, allowing for an abundant, diverse ecosystem and some of the richest farmland on the East Coast.

The Soil

Consider what each soil will bear, and what each refuses.

—Virgil

As a native Long Islander, I can tell you that I didn't know anything about the soil underneath my feet while growing up. I knew it was there when my face hit the ground after a football tackle or how it filled up my baseball uniform sliding into second base. I knew we planted vegetables and flowers in it, that it got all over my clothes and shoes, and that my mother didn't want me to bring it into the house. Back in the day, we just called it dirt.

But soil does a lot more than turn into mud in the rain. Most of us probably don't realize how unique and wonderful our soil is. Like other small wine-growing districts, North Fork soils are exclusive to the region and not found anywhere else in the world. It wasn't until I got to Upstate New York and began studying agronomy that I understood how unique our local soils are. At home on the island, we dug and shoveled anywhere we wanted with abandon; in Upstate New York, it was a chore just trying to put a shovel in the ground. The North Fork, it seemed, had been blessed.

When much of the New World talks about wine growing, there's often a dearth of information regarding soil types and their influence on the wine. This is a distinctly New World philosophy where winemaking techniques often supersede inherent vineyard characteristics. Of course, in the *Cool* New World, we look at things differently; soils can make or break a vintage in cool-climate regions like the North Fork.

Books on Old World wine regions are littered with in-depth descriptions of their soils and their huge effect on wine quality. As a young winemaker listening to wine journalists wax poetically about the famous Kimmeridgian limestone of Chablis or the yellow Tuffeau of the Loire Valley, I realized that little was being said about the soils of Long Island. I decided to learn as much as possible about our local soil and its effects on our wines.

The French have done more work than any other country to study their unique vineyard soils. Their work led to one of the earliest approaches to soil classification, known as the French Soil Reference System (*Référentiel Pédologique Français*). French winegrowers have known longer than anyone else how much influence a particular soil can have on wine quality. In the US, soils are classified by the USDA Soil Taxonomy developed by USDA and the National Cooperative Soil Survey. To understand soils better, scientists typically classify them as consisting of three types of particles—sand, silt, and clay. Sand is the largest particle, while clay is the smallest. Soil types are named for the primary particle size or a combination of the most abundant particle sizes (e.g., "sandy clay" or "silty clay"). A fourth term, loam, is used to describe a roughly equal concentration of sand, silt, and clay and lends to naming even more classifications (e.g., "clay loam" or "silt loam"). As it happens, loam is the predominant soil type of the North Fork.[1]

Unlike the Old World, the soils that make up the North Fork are geologically new, created through the blending action of the Roanoke Point Moraine. As these particles washed down from the top of the moraine, the heaviest sand particles dropped first, creating the underbelly of the region. The lighter silt and clay particles remained closer to the surface and mixed with the remaining sand to form our topsoil. Soil scientists classify them into two main soil associations: *Carver-Plymouth-Riverhead*, which makes up the coastal perimeter along the Sound and Peconic Bay, and *Haven-Riverhead*, which makes up the region's heart. This central area of the North Fork is where most of our agriculture is found, and most vineyards are planted. Haven-Riverhead associations consist of Riverhead sandy loam and, more importantly, Haven loam.[2]

Haven loam is the most important soil of the North Fork and consists of roughly equal proportions of sand, silt, and clay. These soils exist on the outwash plains that slope down from the bluffs along Long Island Sound towards Peconic Bay; they typically lie on top of a bed of sandy gravel and sometimes silt and clay. The soil is perfect for growing fine wine grapes; it's well-drained, limiting the impact of seasonal rain, controlling vine growth, and promoting grape ripening in the fall. Most vineyards on the North Fork are planted on Haven loam and, to a lesser extent, Riverhead sandy loam. These soils provide significant differences in vine growth as well as wine style. Haven loam has more silt and clay and holds moisture longer than the more arid Riverhead sandy loam.

Red wines grown on Haven loam have an expansive mid-palate and can be dense and powerful, while reds grown on Riverhead sandy loams are often lighter in body and color with a pronounced elegance. Both these soils are naturally acidic and contain substantial amounts of aluminum and iron, lending to the minerality and "steeliness" tasters often find in our reds.

A well-known national wine writer once told me she smelled the earth in our wines and was constantly reminded of them while driving through the North Fork in the spring after farmers plowed their fields. Some grape varieties will show their preference for a particular soil, with Merlot and Cabernet Franc growing better in Haven loam. In contrast, more vigorous varieties like Sauvignon Blanc and Cabernet Sauvignon excel on the more arid Riverhead soils.

Unlike many other Old-World regions with undulating topography, there is less diversity above the ground on the relatively level plains of the North Fork. The real action occurs below ground in the subsoil, where sand and gravel dominate in differing proportions, interspersed with pockets of silt and clay. In some places, Haven loam is close to four feet deep; in others, it's barely six inches to the subsoil. Gravel belts can rise to the surface at varying points and are visible in bare-row middles. Clay pockets remain invisible to the naked eye but manifest in vineyards planted over them; vines planted over clay pockets can thrive without irrigation. The more vigorous Bordeaux varieties like Cabernet Sauvignon and Sauvignon Blanc do best on poorer, drier, sandy/gravel subsoils, while Merlot, Malbec, and Petit Verdot reach their highest power on sites with more silt and clay. Cabernet Franc is the exception, as it seems to do well on either type of local subsoil.

Differences in underground soil composition can profoundly influence terroir, creating compelling variations in wine style. Vineyards growing over sandy, gravelly subsoils can still produce excellent wines in vintages with lots of rain. In contrast, in dry years, a predominantly silt or clay layer in the subsoil can hold more moisture, keeping the vines hydrated and healthy, resulting in outstanding quality in these vintages.

The soil of a vineyard is usually not described with the same romantic language as, say, the wines produced from them, but they carry profound importance. The reputation of North Fork wines is rooted in our soil—the result of thousands of years of deposition of particles brought here from as far away as Canada. The confluence of events during the glacial period created unique soils in our region. This place in time—this convergence of water, ice, sun, and rock has left us the ground we live and farm on. No other place in the world can duplicate what we have. On a rare trip to Long Island in February of 2000, the world-renowned Australian viticulturist Dr. Richard Smart stated that the North Fork loams "are among the finest soils for grape growing that I have ever seen in the world." With a benediction like that, we, as winegrowers, are obliged to learn as much as possible about the land on which we farm while simultaneously cherishing and stewarding this precious natural resource—the haven of where we live.

Courtesy of Shutterstock.

The Sea

The sea, once it casts its spell, holds one in its net of wonder forever.

—Jacques Yves Cousteau

An old Bordeaux proverb states, "only vines that overlook the water are capable of producing wines of great quality." I can't say I disagree. Due to the level plains of our topography, most North Fork vineyards don't have a waterfront view. However, our vineyards are closer to the sea than many other wine-producing districts, with most plantings located less than a mile from the shore. I like to say, if you hold up a glass of North Fork wine to your ear, you will hear the ocean.

The power of the sea makes the North Fork such a remarkable wine-growing area. Our cool, maritime region owes its existence to the proximity of the Atlantic Ocean, Long Island Sound, and Peconic Bay, making the area more temperate than many places in the same latitude. The peninsula juts eastward into the sea at an acute angle, the perfect position to capture the blanketing, buffering breezes coming off Long Island Sound. The massive Atlantic, warmed by the Gulf Stream, provides a moderate water temperature throughout the year. This unique orientation into the sea allows the North Fork to receive what few other areas along the East Coast possess—weather systems traveling over a large body of water before reaching landfall. It's even more important when storm systems and weather patterns arrive from the north and west.

Long Island Sound is a tidal estuary of the North Atlantic that ranges in depth from 250 feet at its eastern mouth to forty-five feet near Manhattan. The estuary is home to more than 1,200 invertebrates, 170 species of fish, dozens of migratory birds, and a mix of fresh water from tributaries and salt water from the ocean. Covering over 1,300 square miles, it's ninety miles long and twenty miles at its widest point. Numerous New England rivers drain in the Sound (including the Connecticut and Housatonic), making it less saline than the nearby Atlantic. The Sound can be both surprisingly calm and rough, depending on the weather. At various points throughout the year, the water can be smooth as glass, while at other times, it resembles the Atlantic, with waves ranging from three to five feet in height.[1]

The remains of the Roanoke Point Moraine are seen along the Sound in the towering bluffs along the northern coast of the North Fork. These bluffs are constantly eroded by runoff, waves, and wind, resulting in the narrow strips of shoreline at their base. Sound beaches are covered with rocks, shells, and, more dramatically, large boulders known as *erratics*. These gifts of granite and basalt were left behind as the glacier retreated, dropping the heaviest along the shore like old shipwrecks. Many erratics remain on the beach, mounted in the sand, while others peak out of the water at low tide. Countless more are submerged underwater and hidden underground, discovered only through the work of a backhoe digging a new foundation or swimming pool. Erratics become less numerous towards the Bay.

The average water temperature of the Sound can reach 70°F by the end of the summer, cooling slowly during the fall, helping to push late-season frosts well into November. The average temperature of the winter Sound is 41°F. During the middle of winter, the Sound rarely gets below 32°F, way too cold to swim but warm enough to keep Arctic cold fronts from delivering vine-killing temperatures. On average, the water is at its coldest during the months of February and March—typically between 36°F to 38°F degrees Fahrenheit. By the end of April into May, the average Sound temperature still hovers around 48°F, delaying warm, springtime temperatures and preventing early frosts from damaging newly opened buds.[2]

The more tranquil Peconic Bay, separating the North Fork from the Hamptons, is classified as two distinct bodies of water. The larger, to the west, is called Great Peconic Bay and is no more than thirty feet deep. Little Peconic Bay extends east to open water and reaches depths of over eighty feet. Few erratics exist near Peconic Bay; these beaches are predominantly soft sand with relatively few shells. Due to its shallower depth, the buffering

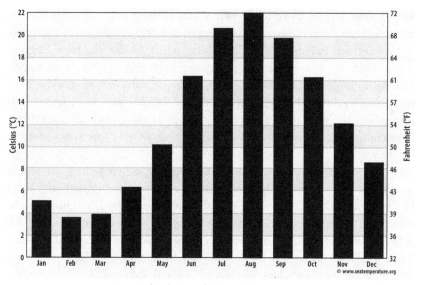

Long Island Sound Water Temperature.

capacity of Peconic Bay is less than the Sound; however, temperatures are further moderated by this calm body of water, especially when winds come from the south. During the summer, temperatures in the Bay can exceed 75°F, while in winter, they rarely drop below 35°F. Even during severe weather, waves on the Bay seldom reach three feet high.[3]

The North Fork is within twelve miles of the Atlantic Ocean, making "Davy Jones' Locker" the region's most significant climate influencer. Along the south shore of the Hamptons, the Atlantic ranges in depth between 100–300 feet until dropping below the shelf break, some 6,000–9,000 feet deep. The ocean provides a tremendous temperature cushion for the entire island. When cold winds come from the east and south, the North Fork is buffered first by the sea, then by Peconic Bay, nestling us in a duvet of temperature protection. The larger land mass of the South Fork provides an additional level of influence, acting essentially as a windbreak to the fierce gusts coming off the Atlantic and capturing the spring and summer fog that rolls off the cold ocean water. I've seen this phenomenon firsthand while working for Mudd Vineyards. Years ago, they maintained a few vineyards across the Bay in the Hamptons. On days when we had to work in those vineyards, we'd leave the warm, sunny skies of the North Fork only to wind up an hour later in a complete fog, chilled to the bone. The Atlantic

Ocean delays the start of the growing season on the South Fork by as much as two weeks due to its cooler water temperature. Of course, the Atlantic can also get angry, generating powerful nor'easters and the occasional fall hurricane that can sweep near our region. In my forty-plus years of making wine, I've only experienced one direct hit from an Atlantic hurricane—the famous 1985 Hurricane Gloria. Knock on wood.

North Fork wines have a deep, seasonal connection between the land and the surrounding sea. Without the incredible influence of these bodies of water, the ability of the region to produce fine wines from *vinifera* grapes would be severely curtailed. It is the sea that allows North Fork wine to be made and the sea that exists in the soul of every glass enjoyed.

A sunset on Long Island Sound. Courtesy of Denise Winter-Peragallo.

The Sun

The Sun, with all those planets revolving around it and dependent on it, can still ripen a bunch of grapes as if it had nothing else in the universe to do.

—Galileo Galilei

It's easy to see why ancient people worshiped the sun. It's the power supply for the earth that keeps all living things warm enough to survive. The sun also provides energy for plants as they use sunlight, carbon dioxide, and water to make glucose—the famous process known as photosynthesis. From the leaves, sugars move into the fruit, promoting ripeness and maturation. Following this relationship, one could say that wine is bottled sunlight. And there's no place where the pastoral blend of sunlight, air, and water all come together as fortuitously as on the North Fork of Long Island.

The North Fork, specifically the village of Cutchogue, has long been touted as the "sunniest spot in New York State."[1] For decades, artists gravitated to the East End because of the area's light and beauty. Many believed the natural light shining on the Bay and the Sound provided reflection not found anywhere else. In the years after the Civil War, artists from New York City explored the North Fork looking for inspiration in the landscape and solitude of the area. The local art scene grew exponentially in the early twentieth century when the Long Island Railroad reached the East End. Artists settled in Peconic and Cutchogue, attracted to the area's sea, woods, and farms, drawn to the soft light reflecting off the water. Farmers would allow the artists to erect easels in their fields and even give them hayrides to the most scenic spots where the sea-stained light filled the open meadows.

North Fork wine pioneer Louisa Hargrave also noted the quality of local light. "I love the water, the sunshine, the farmland, and the light," she said. "Because of the refraction from the sea, you can get a kind of enchanted light." Modern lighting experts have attributed this phenomenon to a "heavy saturation of water molecules in the air," which would catch and hold onto the light. They call this phenomenon "wet light."[2]

Tim Morrin of the National Weather Service once said, "You could argue that the North Fork, in general, is sunnier than the South Fork. The reason for that, meteorologically, is because the South Fork is impacted more by the sea breeze than the North Fork, and the sea breeze often brings in more cloudiness."[3]

All of this begs the question—is this phenomenon real or just a figment of the imagination? While some might be skeptical of these anecdotal observations, good data supports the claims. Sunlight can be challenging to record, but two methods can document regional differences. One way is to refer to the recorded number of sunny days.

By gathering data from local weather stations and news services, we can see how many clear days of sunlight are the norm for various areas in New York. On average, the state has 165 sunny days per year overall. However, when we break it down into specific cities and towns, we get a clearer picture.

In the chart below, the average number of sunny days refers to the number of days in a year when the sky is mostly clear. This includes days when clouds cover up to 30 percent of the sky during daylight hours. The days not counted are mainly overcast, with at least 80 percent cloud cover. All the numbers are annual averages made from years of weather watching.[4]

This and other information show that the sunniest spot in New York is (surprisingly) Central Park in New York City. This figure could be just an artifact of more consistent data collection; however, Cutchogue comes in a solid second place. Central Park receives seventeen more inches of seasonal rain than Cutchogue, which somewhat negates the advantages of the additional sunlight. Looking at this from an agricultural perspective, if we discard the Central Park data and look at the rest of the areas with commercial agriculture taking place on land within their vicinity, Cutchogue comes out on top. Cutchogue shows a higher average number of sunny days (207) than the rest of New York (165) and the entire US, which averages 205 days of sun overall.[5]

Another way to track sunlight is by looking at ultraviolet radiation. The UV index (or UVI) is an international standard measurement of the strength of sunburn-producing ultraviolet radiation primarily utilized to alert people about their potential exposure to the sun. Developed by Canadian scientists in 1992 and adopted by the World Health Organization and World

Meteorological Organization in 1994, UVI calculates the sun's height in the sky, latitude, clouds, altitude, ozone, and ground reflection. UVI essentially represents the amount of sunshine hitting the earth's surface in a given location and ranges numerically from one to thirteen, with one being the lowest level of exposure. Cloud cover can absorb UV light so that cloudy northern areas will have a lower UVI than more southerly regions. UVI is at its peak during summer when the sun is highest in the sky and the days are the longest.[6]

Though it hasn't been around that long, the UVI recorded for the past twenty-five years shows the North Fork with some of the highest levels in New York and New England, especially during August, September, and October. The average UVI for most of Upstate New York is five, while the North Fork and the rest of Long Island are at six. The starkest differences occur during October when the UVI on Long Island is the highest in New York, putting it in the same regional category as Virginia, Idaho, northern Colorado, and southeastern Oregon. On the other hand, Upstate New York is one to two units lower, similar to New England and the Upper Midwest this month.[7]

So does all this mean that Cutchogue is the sunniest spot in New York? Considering the lower precipitation levels on the North Fork, the number of recorded sunny days, and the quality of "wet light," I think I can safely say that the information presented here is quite illuminating.

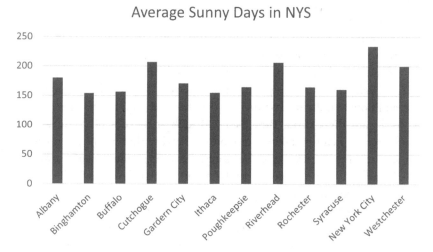

Average sunny days in New York State. Courtesy of http://www.currentresults.combar.

The Aquifer

Water is the driving force of all nature.

—Leonardo da Vinci

Living on the North Fork means we are constantly reminded of water. It surrounds us, protects us, and affects our lives every day. It cools us in the summer, warms us in the winter, and provides endless beauty and enjoyment in so many ways. But many of us don't realize that we're also surrounded by water underneath the ground—a dark, cold river that flows unseen all year long and holds the key to our survival.

Long Island contains three significant bodies of water under the ground—these are our aquifers—geologic remnants from the island's early period of glacial formation. Three separate aquifer layers make up the Long Island system. In sequence from shallowest to deepest, the Long Island aquifers are the Upper Glacial (100–150 feet), the Magothy (500–1000 feet), and the Lloyd (1500–2000 feet). All exist underneath the area of Long Island except for the North Fork, where there is only one—the Upper Glacial. It is the sole source of drinking water for the region.[1]

The Upper Glacial aquifer is replenished from rain and snow. From approximately forty-seven inches of precipitation, about half percolates into the ground and recharges the aquifer.[2] The remaining precipitation evaporates, is used by plants, or runs off into our surrounding waters. In a pristine natural system, the Fork's groundwater would eventually reach the coast and flow into the ocean and Sound while being replenished with new precipitation. But our population constantly requires water to live; the three million people living in Nassau and Suffolk Counties depend entirely on groundwater for all their freshwater needs.

Over 138 billion gallons of water are removed yearly from beneath Long Island. As water recharges the system, it can also contaminate groundwater. Since it is the shallowest and closest to sources of runoff, the Upper Glacial aquifer is the most susceptible to contamination. The Magothy aquifer supplies over 90 percent of the water used in Nassau County and about 50 percent of all water used in Suffolk County. On the North Fork, however, it's all about the Upper Glacial. On average, the North Fork lies ten feet above sea level, making it easy for runoff to vent out to our surrounding waterways. This process can take up to a century to complete, meaning that whatever we do today above the ground can stay with us for at least fifty to one hundred years. This is a critical concept as it relates to sustainable wine growing.[3]

One of the significant sources of groundwater contamination is nitrates. According to the EPA and the Safe Drinking Water Act of 1974, well-testing is done periodically to determine the level of contaminants in drinking water. Years of research have determined that the acceptable level of nitrates in water should be no more than 10 ppm (parts per million). The natural nitrate level in pristine well water from undeveloped tracts of land is almost always less than 3 ppm. Nitrate in North Fork groundwater stems from several different land activities, but overall, the most significant contributor to nitrates is wastewater via septic tank/cesspool systems and home turf grass fertilizers. Development has taken its toll on our water supply.

Agriculture has played a role as well. Over the last century, traditional row-crop farming has increased nitrate levels in many areas. Thirty years ago, many North Fork farmland wells contained nitrate levels of 10 ppm or higher. Many older residues, or "legacy materials," are still found—a remnant of our farming past. Today, only 10 percent of private wells exceed 10 ppm of nitrate, and 29 percent exceed the concentrations of 4 and 6 ppm—a testimony to better farming practices today and greater homeowner awareness of fertilizer use—but clearly, a lot more needs to be done.

For over thirty years, the Suffolk County Department of Health Services (DHS) Groundwater Investigations Unit has been studying the impacts of agriculture on the North Fork, monitoring our groundwater for various chemicals. They were the first such agency in the United States to detect pesticides in groundwater in August 1979. Suffolk County DHS has been studying samples from wells across the North Fork from fallow farmland, housing developments, and traditional agriculture to analyze the impacts on our aquifers. DHS also installed twenty-nine monitoring wells at nine different vineyard sites. Vineyard wells are tested annually, with hundreds of samples taken during the past decade. The result is excellent scientific work providing information to help us become better growers moving forward. [4]

The DHS well-testing data shows that North Fork vineyards are good for our local environment. For one, the average level of nitrates found in vineyard wells was a little over 4 ppm—while the levels found in the traditional agricultural wells were over 13 ppm—three times as high. These results were surprising even to the researchers who (when they began the study) believed internally that vineyard sites would be lucky to reduce nitrates below 6 ppm. Clearly, we have been doing something right in our vineyards. The combination of careful management, permanent cover crops, and limited fertilizer use has resulted in a low environmental impact.[5]

Vineyards have a lower nitrogen impact on our groundwater than a typical house on .5 acres of property (6 ppm) and only slightly higher than a pristine piece of fallow land. It's anticipated that as vineyards continue to thrive, nitrate levels will be reduced even further.

It took thousands of years for our water table to develop. Our well-drained soils and sandy-gravelly subsoils provide the perfect natural filtration system for our groundwater.

When drinking a glass of water from the Upper Glacial aquifer, you taste the melted remnants of an ancient glacier. The truth is that we can

Courtesy of Shutterstock.

live without wine, but we can't live without clean water. It's of utmost importance for local winegrowers to be good stewards of the land—for the local environment and our health and well-being. Consumers of North Fork wines should not only think about the area's beauty but also of the water that runs beneath our feet. It's another part of our terroir—a hidden component yet vital to our lives.

Climate and Weather

Climate is what we expect; the weather is what we get.

—Mark Twain

There's a lot of drama in a vineyard—of course, I am talking about what happens outside. As each day of the growing season begins and ends, another act plays out—just one more part of the ongoing saga of the vintage. To find out what happens, you must stay with the story until the end as there are many ups and downs, twists and turns, enemies and friends—even love affairs. As bit players, we can see the action unfold while we play our part, but we are powerless to control the scenery and the narrative: Mother Nature is the only theatrical producer. Sure, great acting helps, but we need everything to go our way for the production to be a hit. Moreover, vintners only have one performance a year to do it right. Our climate allows us to be making wine here in the first place, but the weather is what determines the greatness of a vintage.

When discussing climate and weather, we're talking about two different things. Climate is the accumulation of weather data—temperature, humidity, sunlight, wind, and other weather conditions that prevail over extensive areas for at least thirty years or more.

In comparison, the weather is what we experience in the short term and describes day-to-day conditions and events. Both consider temperature and precipitation; weather tells us the day's news, while climate is more of a history lesson. From a wine-growing perspective, we often talk about the environment of a small area that differs from the larger surroundings. This is known as a microclimate. In our case, the North Fork has a microclimate that differs from the larger area of Long Island. Microclimate is arguably the single most important factor determining the character of wine from any given district and largely determines which grape varieties are viable. Our microclimate on the North Fork allowed us to become one of the East Coast's largest producers of European wine grapes.

The North Fork is considered a cool maritime climate for obvious reasons—we're located in a northern latitude, near the sea. Take away the surrounding water, and the North Fork climate would probably be like the Lower Hudson Valley or southern Connecticut. Water drives our climate; without it, we would not exist as a wine region. Our maritime climate has allowed us to grow the European grape varieties that were in the dreams of the early colonists and still elude many growers on the mainland today. The East End remains the only wine district on the East Coast with no significant commercial plantings of native or hybrid varieties of wine grapes.

As mentioned previously, the moderating properties of the sea can be profound, providing buffering breezes and tempering frost risk, effectively counterbalancing extreme temperatures. Our climate has warm (but not hot) summers and cool (but not cold) winters—necessary conditions for a winegrower. The deep waters offshore take months to warm in the summer, holding temperatures long after the surrounding coast has cooled down, helping moderate the climate well into the fall. We have one of the longest growing seasons in the Northeast, with beautiful ripening conditions consisting of warm days and cool evenings.

Unlike other regions in the New World, the North Fork benefits from seasonal rainfall. The Northeast has no water shortage issues as our precipitation provides plenty for our plants to grow. While the rest of Long Island records an average of forty-two inches of rain per growing season, the North Fork is much drier, averaging about thirty-two inches of rain during the same period. The contrast between neighboring areas is even more profound, with Connecticut, New Jersey, and Downstate New York accumulating between thirteen to fourteen more inches of rain per season. That comes to an average of two to three more months of growing season rainfall for these areas compared to the North Fork.[1]

It's well known viticulturally that rainfall and the amount accumulated can significantly affect vine growth and fruit quality. More rain traditionally means a lower quality crop for winegrowers; however, the North Fork's well-drained loam and gravelly subsoils have allowed for good vintages even in rainy seasons. Unlike other regions in the East, very little standing water can be seen on the ground after a North Fork rain. Recent studies have shown that in the near future, effects from rain during the ripening season may be offset by increases in temperature from climate change.

Ironically, the drier conditions on the North Fork are attributed to the surrounding water. Precipitation fronts and summer thunderstorms are often diverted to the north and west during the growing season due to the prevailing southwest winds blowing over the North Fork. It's not uncommon for New Jersey and Connecticut to be experiencing heavy downpours during the growing season while the North Fork remains dry and sunny.

Occasionally, we can have a few weeks during the summer when we can approach drought-like conditions; the need for irrigation depends on the soil and subsoil of a vineyard. When irrigation becomes necessary, drip application is the preferred method.

I drilled deeper into the data and compared precipitation levels and timing to the quality of past vintages. The most outstanding North Fork vintages occur when the combined rainfall in August, September, and October is less than 10 inches. This data held true through the extraordinary vintages of 1988, 1993, 2007, 2010, 2019, and 2022. Regardless of what happens earlier in the season, rainfall levels during August, September, and October ultimately determine wine quality.

The wind coming off the water does more than just provide much-needed buffering—it can also significantly help control mildew and other grape diseases. Compared to the world's more famous wine-growing regions, the North Fork is the windiest in the Western Hemisphere, with an average wind speed of 9.1 mph. Only Tours in the Loire came close at 8.9 mph. During the fall months, the North Fork outdistances the competition with an average wind speed of 11 mph, with Tours at 9 mph. Most other wine-growing regions on the West Coast and Europe have calmer average wind speeds. Only New Zealand is windier than the North Fork.[2]

Although the amount of sunshine and rainfall can affect the length and quality of a vintage, an essential factor in determining the commercial viability of any agricultural area is the number of days between the spring and fall frosts. This is the period between the spring's last frost (32°F) and the first frost (32°F) in the fall. This period is also known as the *growing season* since grapes will not continue to grow once temperatures drop to 32°F. For the last decade, North Fork farmers have been blessed with an average growing season of 219 days.[3]

It's well known in the Old World that areas adjacent to large bodies of water are less subject to frost than areas farther removed. Spring temperatures on the Fork delay plant growth, reducing the chance of frost damage. Areas affected by warmer continental climates, such as Connecticut, Downstate New York, and New Jersey, experience earlier plant growth in the spring, increasing the likelihood of frost damage.

In analyzing the frost-free data, it becomes clear that the influence of the water on the North Fork becomes even more significant. A 219-day North Fork growing season is more than thirty-nine days longer than New Jersey and Westchester/Downstate New York and fifty-nine days longer than the Connecticut average. The North Fork can have as much as four to six weeks more growing season than the surrounding mainland; the most significant climate factor separating the North Fork from the surrounding land masses highlights the power and energy of the surrounding water.

From 1960 through 2021, the North Fork accumulated an average of 205 frost-free days. Since 2000, the season average has increased to 216 days—almost three weeks longer. For the last ten years (since 2011), we've been averaging 219 days of the growing season, a little more than three weeks longer than the region's first four decades of modern wine growing (see appendix).[4]

Compared to the mainland regions to the north and west, the North Fork is slightly cooler during the hottest summer months and significantly warmer during the coldest months of winter. In all seasons, there are few extremes in either direction of the thermometer, making the North Fork an ideal location for growing European wine grapes.

The Seasons

SPRING

In the Spring, I have counted 136 different kinds of weather inside of 24 hours.

—Mark Twain

Spring is the time for new beginnings and the official start of a new vintage. A typical North Fork spring starts off cold, as water temperatures in early March are still at their lowest point. Spring is also the time for rain, charging the soil and replenishing groundwater for the coming summer. It doesn't start to warm up until the beginning of June; when it does, it can happen quickly. As Dave Mudd famously exclaimed about the North Fork spring, "You go from wearing long underwear to shorts in one day!" Mark Twain must have been talking about the North Fork.

Vines in the spring seem to understand there's a limited amount of time for them to do their thing. Grapes are notoriously slow to start growing in the spring. In early May, when the grass starts growing, flowers bloom, and trees leaf out, grapevines are just beginning to wake up. When visiting the vineyards in spring, I've had quite a few people ask if our vines were dead. I assured them they would grow like mad and cover the trellis in a few weeks. Young shoots can grow as much as an inch per day in late spring and early summer. A month later, trellises that looked bare can be completely covered in foliage. It's good that the vines aren't fooled too easily by warm April days, as this delay in growth provides much-needed protection from late spring frosts.

The average date of our last frost is April 25. Historically, it was rare to have bud break before this date; however, we are seeing warmer spring temperatures and earlier growth with climate change. Occasionally we can have frosts in early May, but they rarely happen after the buds have sprouted. Hopefully, climate change doesn't affect these conditions too dramatically.

By early March, the ocean has cooled significantly from the low winter temperatures. Breezes coming from the south and west at this time of year pass over the cool waters, reducing air temperature as it passes over the North Fork. Because of this maritime effect, the Fork has long, cool springs, with many plants showing delayed growth compared to their more inland counterparts. All one must do to see this effect in action is to drive west toward New York City; it can be like traveling from winter into spring in under two hours. This is a significant advantage for grapevines as cooler spring temperatures delay the onset of growth and reduce the likelihood of an economically devastating spring freeze.

Towards the end of spring, usually around the middle of June, the vine shoots are a few feet long, and the small, yellow grape cluster flowers open and bloom. Grapevine flowers aren't much to look at as they don't have any petals, but they have a subtle, delicious smell that pervades the entire vineyard. Bloom is the time when the quantity of the crop is determined; cold and rainy weather can result in fewer flowers pollinated, while clear, calm, sunny days can result in a larger number of berries on the clusters. Bees and other native pollinators are attracted to these rudimentary flowers but aren't really necessary as all grapevines are self-pollinated. Still, the vines are buzzing with life during this period, and pollinator activity is welcomed. In less than two weeks, bloom is over, the flower parts detach, and the small, hard green berries start their journey towards ripening.

Spring is also the time when new wines get bottled. I like to have white and rosé wines ready for bottling before the summer heat moves in. It's also when we start racking (moving wines off of their sediment) red wines that have been in barrels since the previous fall.

During my first spring as a winemaker in Bridgehampton, I was expecting my first delivery of empty glass bottles. As the company didn't own a forklift, I had no way to unload the one thousand empty cases that were arriving. I called my friend Connie Kalish, a retired potato farmer, who said he'd make a call. "Just give the guy a case of wine, and you're all set," he told me.

Right after the truck of bottles pulled up, an old giant farm tractor with a potato loader came roaring down the driveway. On the bucket were two flat pieces of steel held on by some rusty chains. The farmhand drove it up, parked it, and jumped down.

"Connie said you needed to unload something," the farmhand said.

"Yes, thanks so much," I replied. "The truck's right over there."

"Oh, I can't stay," the man said. "I'll pick it up later. You got that case of wine?"

So began my first experience unloading pallets of glass bottles. It's stressful enough to do with a proper forklift, but the contraption I had to use put the fear of God into me. Luckily, I had an empathetic truck driver who guided me every step of the way. Of course, it was in his self-interest as he wanted to escape the Hamptons's traffic before it got too late. I cranked the beast to life and began operating the controls. The machine had about as much finesse as a bull in heat. Eight-foot-high pallets stacked with lop-sided cases of empty bottles, all held together with a single piece of nylon cord. I didn't break a single bottle, but I needed a drink when I was done.

All the seasons on the North Fork have their sounds, sights, and smells. Here are the things that flow through my senses and signify to me the changing seasons of the year.

Sounds: spring peepers, northern cardinals, Carolina wrens, lawnmowers.

Sights: tractors plowing, daffodils, forsythia flowers.

Smells: newly tilled earth, lilacs, freshly cut grass, strawberries, rhubarb.

SUMMER

Shall I compare thee to a summer's day? Thou art more lovely and more temperate: Rough winds do shake the darling buds of May, And summer's lease hath all too short a date.

—William Shakespeare

Summer on the North Fork is, without question, the best season of the year. It's the time when the area shows off all its beauty. The weather can be near perfect with the bright, clear light and the sea warm enough to swim comfortably. Summer is also when local farmers begin to show off the region's delicious bounty.

During the summer, breezes from the water keep the average temperatures on the Fork slightly lower than its more inland neighbors. One need only go to the seashore on a hot, humid day to realize the incredible cooling effects of ocean breezes in the summertime. Since plants and grapevines do not conduct photosynthesis at temperatures greater than 90°F, these moderating winds can benefit healthy vine growth.

Historically, the warmest month of the year on the Fork is July, with an average high of 79°F and a low of 66°F. Water temperatures are still relatively cool at the beginning of the month but warm up quickly when the hot days roll in. Even so, the Fork experiences very few days over 90°F. Our long, warm summers are tempered by cooling breezes off the Sound and the Atlantic Ocean that prevent excessive summer heat. In Cutchogue, there's an average of four days annually when the high temperature is over 90°F, which is cooler than most places in New York; temperatures in Manhattan are typically 10°F–15°F warmer in summer than in Cutchogue.[5]

Work in the vineyard during the summer is in a frenzy, with leaf pulling, fruit thinning, grass mowing, weed hoeing, and disease control taking up the vineyard manager's attention. Careful consideration is given to vine health and water needs, with irrigation applied if drought stress becomes too much. It's the busiest time for the vineyard, and it's when the vines and their crop are readied for the coming ripening season.

For young and newly planted vines, summer was historically a difficult time as deer and rabbits would feast on the tender new shoots. In the old days, conscientious growers used open milk cartons or chicken-wire cages to keep animals away. These makeshift solutions were eventually replaced by the widespread use of grow tubes—plastic cylinders that allow light to penetrate and are reused for many years on new vines.

Working in a vineyard during the summer is no walk in the park. It can be hot, humid, and monotonous, with each task repeated hundreds of times as each vine needs individual attention. For new vineyards, it's often the time when the trellis gets installed. Posts are laid out, pounded, and anchored, wire nails hammered in, and thousands of feet of high-tensile wire stretched out, attached, and tightened.

A big part of summer work in the vineyard is weed control and canopy management. For the first ten to fifteen years of modern North Fork viticulture, most growers clean-cultivated the entire vineyard floor using a disc alongside a side-mounted contraption called a grape hoe. A summer day spent driving these implements down vineyard rows would result in the tractor driver being covered entirely in dust by the time they finished. As we learned more about soil erosion and vine health, growers moved to a permanent sod or grass cover crop between the row middles.

Canopy management has also evolved. Back in the day, workers would lift individual vine shoots and tie them to the wires with rope or twist ties—a tedious, slow job. When I first started, the latest technology for tying vines was called the Max-tapener, a hand-held staple-gun-looking device that held rolls of green ribbon, which you would pull around both shoot and wire. When you clamped the machine and closed the latch, it stapled the ribbon together and then cut it. To be kind, it was not a time saver.

Other growers would just let the vines grow into the row middles, creating a jungle of foliage. Driving a tractor through this mess was another adventure, especially before the days of closed tractor cabs. I'd get continually slapped in the face with vines while driving down a row, tearing away and breaking green shoots that wouldn't give as easily. Forward-thinking growers like Kip Bedell championed movable catch wires, which could simultaneously lift the shoots of many vines, accomplishing the task in much less time. This technique has become a standard operating practice for North Fork vineyards and is part of the training system known as Vertical Shoot Position or VSP. It keeps shoots vertically positioned on the trellis while helping to expose the fruit to sunlight.

During the middle of summer, work in the cellar enters its slowest phase. Whites and rosés are already bottled, and the reds are resting in barrels, slowly softening tannins and developing flavor. By the end of summer, planning and organization for the coming harvest become the top priority. Summer in the vineyard comes to a psychological end during the middle of August when fruit truly begins to ripen, and the weather becomes even more critical for success.

Sounds: children laughing in the surf, seagulls squawking, ospreys chirping, cicadas buzzing.

Sights: butterflies dancing, fireflies and carnival lights at night, deer grazing in open meadows.

Smells: rain on the ground, peaches, corn, tomatoes, saltwater spray.

FALL

Autumn, the year's last, loveliest smile.

—William Cullen Bryant

Fall is showtime for both vineyard and cellar. The actors have prepared all spring and summer long for the final act. As our performance is an outdoor event, all we can do is hope that the weather cooperates and that we can put on the best show possible. Fall is when the maritime influence on the vintage takes full effect. As summer wanes, the temperature of the Sound and Bay are at their highest, extending the season by preventing early frosts and allowing us ample time to ripen our fruit well into October and even November. For us, the summer eases into fall with hardly a sigh. Early September days can be some of the most beautiful on the North Fork. As the end of the month closes in and October takes over, fall soon reminds us of what's coming as the nights begin to cool down and the sun just doesn't feel as strong. This makes for near-perfect ripening conditions for European wine grapes.

The beginning of the ripening season starts with the period known as *veraison*—a French term that describes the time when grapes turn color from green to either yellow/gold, copper/pink, or blue/purple/black depending on the variety. Its literal definition is a "change of color of the grape berries." Although this usually takes place sometime in the middle of August, it's the unofficial beginning of fall and the time when winemakers must start paying attention to the speed and timing of fruit maturity.

When *veraison* begins, and the grapes start to show the rest of the world what they are, the birds start to take an intense interest. Starlings, grackles, robins, and blackbirds combine to make up flocks that can number in the tens of thousands. Dark clouds of birds are a wonder to behold, swirling and twisting into different shapes as they dive and swoop over the tops of the vines. Native North Forkers will tell you the sound from these large flocks reminds them of an old squeaky potato harvester chattering over the fields.

Without deterrence, birds will wipe out the fruit from an entire vineyard. Growers during the early days used to mitigate this problem by picking early to avoid damage. This was a particularly sensitive subject, especially if you were buying fruit from another vineyard. The choice was often between buying less-than-perfect fruit or no fruit at all. While this strategy worked to preserve yields, it resulted in some less-than-ripe flavors and lower quality in some early North Fork red wines. I know. I made a lot of them.

Back in the day, before bird netting, we used all kinds of crazy techniques to try to keep the marauding creatures away. Slapping two pieces of wood together, plastic owls, shotguns, mylar tape, sonic alarms, and the bane of everyone living near a vineyard—the propane cannon. All these things worked for a short time, at least. But as soon as the noise or movement stopped, the birds, who would roost in nearby trees, would quickly return to do their dirty work. As satisfying as some of these deterrents were to use, they were ultimately a waste of time and money.

One of my favorite wastes of money in this regard was a 1969 VW beetle I purchased for fifty dollars from the side of the road. It had no key, so I jammed an old corkscrew into the ignition to start it up. It would never pass inspection, but it fit nicely between vine rows. I bolted a propane cannon to the roof and ran the gas line into a small BBQ tank in the back seat. I drove that old jalopy around for hours, honking the horn while wearing headphones since the car's interior boomed like a firecracker inside a dumpster. The birds disappeared while I was driving, but as soon as I parked and got out of the Bug, I could see them flying back in. It was immensely frustrating.

Thank God a few insightful growers started looking into netting as a sustainable solution to protect their crops. It was expensive, but so were the lost grapes. The first bird netting research was done at the Cornell University Long Island Horticultural Research and Extension Center in Riverhead—LIHREC for short, by Larry Perrine, the first wine grape researcher on Long Island. His initial work, which dates to 1987, showed that netting was the most effective tool against bird damage and economically sustainable with a high-value crop like wine grapes. To this day, bird netting has become a standard operating procedure for successful vineyard management in our region. Not only does the netting reduce damage and loss, but it also allows vintners to hang fruit on the vine for as long as possible, increasing ripeness, aromatics, and tannin development in the finished wines. There is no doubt that this work and the resulting implementation of bird netting was the single most important development in increasing local wine quality since our modern industry began.

The harvest typically lasts through September and October. Occasionally we'll pick Petit Verdot or Cabernet Sauvignon in early November, but this is becoming rarer with climate change. I've started many a vintage by picking Chardonnay in sunny, hot September weather, only to finish up the harvest with a final picking of Petit Verdot during a November snow squall. Only after all the fruit is in the cellar can the vineyard crew breathe a sigh of relief and begin to get some much-needed rest. For the cellar, however, the work is just beginning.

Sights: tractors carrying grapes down the road, wild turkeys grazing, leaves changing color.

Sounds: Geese honking, dry grass bristling in the wind, squeaking flocks of starlings, the rattling of a mechanical harvester.

Smells: dry leaves, freshly picked grapes, fermenting juice.

WINTER

To appreciate the beauty of a snowflake it is necessary to stand out in the cold.

—Aristotle

Winter is the season when most other types of farming come to standstill. Years ago, it was the time when local potato farmers would travel to Florida to visit friends and relatives. No such break takes place in the vineyard or the cellar. Vineyard work continues all winter long, starting in December with pruning. In the cellar, winter is when new wines are monitored through tasting and lab analysis, and blending trials begin.

Low winter temperatures have historically been a main limiting factor in planting European grapevines in the Eastern United States. The North Fork largely became a wine district due to the area's mild winters. During the coldest months of the year, the Sound, Bay, and Atlantic all have tremendous buffering effects due to their accumulation of heat from the summer and fall. Winds from the southwest are warmed by the Atlantic Ocean, raising the temperature of the Fork as it passes over us. These breezes and those coming off the Long Island Sound work to keep winter minimum temperatures high enough to prevent commercial vine damage. Rarely do we see temperatures below zero, and the area is consistently warmer than any of the areas surrounding us. The maritime effect is not realized in New Jersey or the rest of New York State as these land masses lie downwind from the prevailing breezes and receive colder continental winds traveling overland from the west.

The Fork's average annual low temperature is 43.5°F, 2.5°F warmer than Westchester, Downstate New York, and New Jersey, and over 4°F warmer on average than Connecticut. Westchester County averages 1.2–3.0 days annually when the nighttime low temperature falls below 0°F. From Riverhead to Cutchogue, there are less than 0.3 days annually when the nighttime low temperature falls below 0°F. Cutchogue averages 0.2 days annually below 0°F, and is warmer than any other place in New York other than Manhattan.[6]

Grapevines require a certain chilling period to remain healthy and renew themselves for the following year. They don't completely shut down but undergo a period of dormancy. Just as a tree loses its leaves in the fall, so does a grapevine slowly go into hibernation. Winter is the time for vines to rest; like people, they will eventually fail if they don't get enough sleep. Most grapes cultivated in the tropics never stop growing and eventually become stressed out and die. Every vine needs a period of cold weather to break its dormant period and begin growth anew again. Grapes are notoriously deep sleepers, and it takes a lot to wake them up. Only when spring temperatures reach a minimum average of 50°F and the day length starts to increase will a vine wake up from its long winter's nap. We don't usually see enough of those temperatures until late April.

Years ago, the belief was that any site on Long Island would be suitable for wine grapes. The winter of 1984, however, proved that the theory was incorrect.

I still remember the night. It was late January, and I had just started pruning the week before. The weather service predicted the forecast to be very cold, with lows approaching 0°F. I was concerned, yet most *vinifera* vines are still ok at those temperatures. When my wife and I went to bed, the thermometer outside my window was in the single digits. It was a crystal-clear evening without any wind, and the snow on the ground sparkled under the light of a full moon. It was a beautiful yet ominous evening.

By early morning, the minimum temperature recorded showed −12°F. I felt sick to my stomach as the phone rang. My old boss Dave Mudd had already heard about the low temperatures. "You better get out there and cut some buds and see if the vines are alive," he said. "And stop any pruning if you've already started."

I went out, took some cuttings, and brought them inside the house to warm up, arranging them in a vase as if they were flowers. Our consultant at the time, the highly regarded vintner from the Finger Lakes, Hermann Wiemer, came down to visit me the next day. We sat at our farmhouse table with small razor blades and began slicing open the buds to look at them. "Dead," he said after each bud slice. "Dead," as he kept cutting. "This reminds me of the vineyards in Germany after the war," he added, shaking his head. The vineyard I had worked so hard to whip back into shape all season was essentially devastated. So was I.

As it turned out, the vineyard site in Bridgehampton was way too low and was a sinkhole for all the cold air surrounding the property. The depression in the land wasn't very dramatic, but cold air doesn't need much of a slope to move downward. The result was temperatures well below 0°F in the vineyard—and vines killed right down to the ground. I didn't have to prune anymore.

It's incredible how resilient vines are. The following season, new shoots started growing up from the base, curling around the brown, dried skeletons from the year before. I was able to rebuild the vines and get them back in production for another crop in 1985—only to be thwarted again—this time by Hurricane Gloria—the last major hurricane to hit Long Island directly. It didn't matter much because the vineyard experienced another major freeze two years after that, and both Lyle and I abandoned the idea that we could grow grapes on the site. It was just not sustainable for wine grape production.

The vines I've planted and worked with on the North Fork have all been on much better vineyard sites with good air drainage and elevation away from damaging temperatures. But living through a couple of winter vineyard massacres profoundly affects you. When cold spells arrive now, I still get anxious and worried—like a knee acting up during rain. This experience taught me (and others) the importance of site selection for any vineyard. It made me keenly aware of the damage that can occur if one is not truly careful about where their vineyard is planted. Proper site selection for a vineyard is one of the most (if not THE most) essential decisions a prospective winegrower can make. Failure of due diligence in this regard can result in pouring money down the drain.

The result of my experience in Bridgehampton led to a more thorough understanding of our local terroir. Gone was the notion that any piece of land on Long Island was suitable for growing European wine grapes. It became clear that even in our moderate and forgiving climate, there were certain areas where wine grapes would not grow. Good site selection requires a complete topographical analysis of the land. Low spots in the land must be avoided, as even the slightest downhill slope provides a runway for cold air. The best vineyard sites are like an inverted bowl, allowing cold air to flow away from the vines in every direction, keeping them safe from injury. I hope my experience will help other growers make better site selection decisions in the future.

Sights: snow falling on the vineyard, pruning crews bundled up against the cold, holiday lights.

Sounds: blowing wind, the roar of the seashore, melting icicles dripping.

Smells: wood fires, newly racked wines, the scent of snow in the air, pine and cedar trees.

Growing Degree Days

A change in the weather is sufficient to recreate the world and ourselves.

—Marcel Proust

One of the main tools used to track climate in agriculture is Growing Degree Days, or, as they are famously known, GDD. GDD measures the amount of heat accumulated during the growing season; for grapes in our neck of the woods, this is calculated from March 1–October 31. It's essentially the average daily temperature minus a base temperature of 50°F, which is the temperature where growth starts for woody plants in the Northeast. GDD calculates the days necessary for plants and other organisms to complete their growth and development. Winegrowers and other farmers use this system to determine the boundaries of climate, helping to predict the quality and timing of the season.[1]

Cooler regions have fewer GDD, and warmer regions have more GDD. A. J. Winkler and Maynard Amerine from University of California, Davis developed the system beginning in the 1940s. The Winkler scale has since become widely used to classify wine-growing regions worldwide.

Winkler's climate classification of regions, from coldest to hottest:

- **Region I** (2,500 degree-days or fewer)

- **Region II** (2,501–3,000 degree-days)

- **Region III** (3,001–3,500 degree-days)

- **Region IV** (3,501–4,000 degree-days)

- **Region V** (more than 4,000 degree-days)

The average GDD on the North Fork since calculations began in 1940 is 3,100 days. During the first part of this period, from 1940–1970, the average GDD for the North Fork was 2,932. During the decade of the 1970s it was 2,987, and from 1980 until 2001, the average GDD climbed to a level of 3,252. From 2010 through 2021, our average GDD rose to 3,510. GDD accumulation in 2010 was the highest ever recorded on the North Fork—3,762 days. These figures show that climate change is alive and well; we can see the region moving from Region II to a solid Region III and occasionally approaching Region IV levels in warm years.[2]

Is this all part of the global climate change trend? I believe it is, as do many climatologists and other scientists who have spent their lives researching this phenomenon. Some will disagree, but as Daniel Patrick Moynihan once stated, "Everyone is entitled to their own opinions, but they are not entitled to their own facts." And the facts speak for themselves.[3]

Generally speaking, fine wine production is limited to Regions I–IV. These climates allow for more desired levels of sugar and acid balance in the grapes, as well as a slower rate of fruit maturation, preserving aromatics and ripening tannins. One can also note that variations of as few as 100-degree days can separate areas with different climatic conditions.

The North Fork used to be categorized near the high end of Region II; nowadays, we can safely say that we are near the high end of Region III. No other area in the Northeast even comes close. The North Fork historically has an average of 166 more degree-days than Westchester/Downstate New York and 324 more degree-days than Connecticut. Connecticut, on average, is a borderline Region II, with some years having Region I conditions. On the other hand, New Jersey has considerably more degree days (as many as 306 days) than Long Island. New Jersey is classified as a Region III, with some locations approaching Region IV status in warmer years. The closest in terms of GDD to Long Island is the area of Westchester, also a Region II. However, as one will see from the other data presented, there are much more significant differences between these two areas.[4]

Some folks like to use GDD alone as a vintage qualifier, but GDD is just one tool to help us understand our climate and weather as it unfolds throughout the season. It can tell us where we are in heat accumulation, but as we've seen, it doesn't tell the whole story until the vintage is safely in the tanks. Rain, cloud cover, humidity, and poor vineyard management can all work to counteract the effects of high GDD. I've seen seasons with high GDD produce average wines due to excessive late-season rains. Conversely, lower GDD years can turn out great if rainfall is minimal.

GDD tells us the limitations and potential of our climate for growing crops. It can help vintners decide what varieties to grow and what kinds of wines they can expect to make. Most importantly, it can also tell us what our climate has been like in the past and where it will go in the future. Over the past forty years, we've seen a slow but steady increase in heat accumulation—another clear indication of the effects of climate change.

The North Fork vs. the Hamptons

Geologists have a saying—rocks remember.

—Neil Armstrong

While some people across the country often confuse the North Fork with the Hamptons, the two areas could not be more different. Once a farming and fishing area, the Hamptons became a popular summer getaway for artists, celebrities, and well-heeled New Yorkers looking to escape Manhattan. Blessed with miles of beautiful ocean beaches and a mild climate, it slowly morphed into a cluster of affluent villages and remains quite posh, with high-end restaurants and clubs alongside a trendy, beach-party culture. Many of the rich and famous spend their summers on spacious estates nestled along the Atlantic Ocean, often hidden behind tall hedgerows. Name a famous person, and they probably own a house there.

Across the Bay, the North Fork remains more secluded from Manhattan and has maintained its rural and agricultural heritage for centuries. Located across the Sound from Connecticut and without Atlantic Ocean beaches, the North Fork developed a closer cultural relationship with New England than Manhattan. Forty years ago, most people in NYC had never heard of the North Fork and didn't know it existed, but they probably ate many of the fruit and vegetables grown from the area.

Over the past few decades, as the North Fork entered New York's consciousness and the fishing industry waned, the older connection to New England began to fade. If the Hamptons is associated with the wealth and glamour of Manhattan, one could say the North Fork now has its ideological connection with Brooklyn. This analogy has become more evident with the recent "discovery" of the North Fork during the COVID-19 pandemic. The North Fork is more rural, grittier, down-to-earth, and less tony than the Hamptons. North Fork beaches, while beautiful and serene, are rockier and don't have any waves for surfing. It tends to be more middle-class and politically conservative, and if we have any famous people living here, no one knows about them. It's preferred by many for its small-town vibe and more affordable real estate. While we're becoming trendier and attracting tourists like never before, the North Fork will always retain its rural, agricultural zeitgeist. We are, thankfully, never going to be like the Hamptons.

There are many reasons why the North Fork developed into the main wine-growing region on Long Island instead of the Hamptons, but the initial driving factor revolved around economics. While the land mass of the Hamptons is almost double the size of the North Fork, the price of land has historically been more than four times as expensive. This price difference existed during the late '70s and early '80s when the first plantings went into the ground and continues to this day.

More importantly, the two regions' climates, weather, and soil differ significantly. This became evident through the American Viticultural Area (AVA) applications that I authored. I researched the data to achieve approval for both areas during this process, and you can view the North Fork AVA application in its original form in the appendix.

As mentioned previously, two separate geological events created the landmass of Long Island. The terrain of the Hamptons resulted from the first glacial deposits formed by the Ronkonkoma Moraine 55,000 years ago. The North Fork was shaped more than 30,000 years later during the retreat of the Roanoke Point Moraine. Both glacial events left diverse geological deposits resulting in loam and sandy-loam topsoils on the North Fork and heavier silt-loam soils on the South Fork. Hampton's soils can therefore hold more water and need less irrigation than North Fork soils.

Precipitation is slightly higher on the South Fork, with a mean average annual rainfall of 51.5 inches compared to 47 inches on the North Fork.[1] This difference of four inches of rain also holds true during the growing seasons between March-October for these two regions. Due to the confluence of slightly higher precipitation and heavier soils, farmers on the South Fork rarely irrigated their plantings of potatoes and strawberries, while their northern neighbors could not grow these crops without additional water.

The most significant differences are about temperature. Climatically, the North Fork is warmer and has a longer growing season than the Hamptons. Weather data from 1988 through 2021 show an average number of Growing Degree Days of 3,361 for the North Fork and 2,904 for the Hamptons. This significant difference of 457 days amounts to almost a month's worth of growing season. Most of this is due to the overwhelming influence of the Atlantic Ocean on the South Fork, especially in early spring when cold air from the sea delays the onset of warm weather.[2]

In addition, the number of frost-free days drastically diverges between the Forks. From 1960 through 2021, the North Fork accumulated an average of 205 frost-free days, while the Hamptons had 182 during the same period—a difference of almost one month. The two regions had similar growing seasons in the 1960s and '70s; the North Fork started to warm significantly in the late '90s and early 2000s. GDD and frost-free data show that the South Fork got significantly colder in the 1980s, a period that ran through the end of the 1990s. Overall, the length of the South Fork growing season seems little affected by climate change as the data is more or less statistically the same for the past forty years. Again, this is probably due to the proximity and overwhelming influence of the Atlantic Ocean. Recent data, however, shows some of the warmer years on record for the region, so it's fair to say the Hamptons will also see some warming.[3]

During the last ten years, the North Fork experienced an average increase of around twenty days of growing season due to climate change; there remains approximately a thirty-day difference in frost-free days between these two growing areas. This distinction undoubtedly affects the types of wines produced. While the North Fork has based its reputation on producing excellent red wines from Bordeaux varieties, growing high-quality reds on the South Fork from these grapes has proved to be more challenging. The Hamptons remains an area focused predominantly on producing beautiful white and rosé wines due to its cooler climate and heavier soils.

While climate change has affected conditions in both areas, significant differences between these two regions remain. Whether the issue is climate, geology, culture, or economics, the North Fork and the Hamptons, while being close neighbors, will always be different worlds.

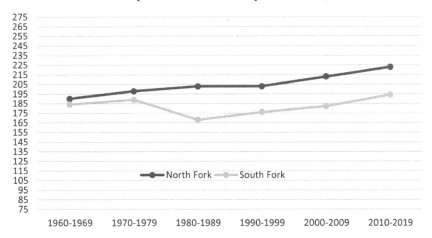

History of Frost Free Days 1960-2019

Courtesy of CCE Suffolk County.

Year	North Fork	South Fork
1960–1969	190	184
1970–1979	198	189
1980–1989	203	168
1990–1999	203	176
2000–2009	213	182
2010–2019	223	194

Terroir

We are born at a given moment, in a given place and, like vintage years of wine, we have the qualities of the year and of the season of which we are born.

—Carl Jung

The sommelier pours the wine into a glass. Lifting it by the stem, she holds it up to the light admiring its color. With a twist of her wrist, the glass moves, and the wine spins around like a ballerina. She tilts the glass, raises it to her nose, and sniffs deeply. She drops the glass slightly, swirls it again, and brings it to her lips.

"I get a distinct saline-minerality along with white flowers and a little bit of grass and seashell," she says. "It's delicious. Where's this from?"

"The North Fork," I tell her. "It's what our terroir does."

Take a walk along any North Fork beach, and you'll inevitably find a whelk shell, just like the ones native Corchaugs once used to craft wampum. Hold it up against your ear, and you'll hear the sound of the region. Hold it up to your nose, and you'll smell the nuance of the sea, which is the common thread running throughout all of our wines. This is our terroir.

Terroir—the heart and soul of winemaking—remains an elusive concept to explain. As a French word, it has no English counterpart. The French define terroir as the relationship between the characteristics of an agricultural product and geographic origin that might influence it. It's commonly described as the "taste of a place"—the flavor that wine "absorbs" from the climate, soil, and location where it is grown. I believe terroir is the overriding influence of a wine region; it transcends specific grapes, showing itself consistently across a broad varietal palate. Terroir is truly a wonder of our existence; how this phenomenon happens has been the subject of some debate.

The concept of local flavor in wine was first noted by Cistercian monks working in the Roman-planted vineyards of Burgundy. Over time, the understanding of terroir became so ingrained and omnipresent to wine producers that it led to the creation of the Appellation d'Origine Contrôlée (AOC) program in France. The French AOC system defines more than 360 geographically distinct areas and stipulates certain grape varieties and winemaking practices for each separate appellation. The system led to the creation of the Denominazione di Origine Controllata (DOC) and Denominazione di Origine Controllata e Garantita (DOCG) systems in Italy and the Denominación de Origen Protegida (DOP) in Spain. It also influenced the development of other wine classification systems around the world, including the American AVA program. The heart of all these programs lies in the belief that terroir—the effect of geography, location, climate, and soil on grapes, is the most important factor in determining a wine's character and quality.[1]

Despite this long history, some still believe terroir is a myth—marketing nonsense dreamed up by French and European winemakers to advance their products. Others insist that terroir is a religion and cannot be proved by science. Conversely, a few radical adherents of terroir believe the minerals found in the soil are directly brought up through the vine and into the grapes, expressing these characteristics in the wines they make. Over the past decade, these misconceptions have been debunked, and new research has drilled even deeper into the concept to help us better understand what's going on.

Recent work done at the University of California, Davis found evidence that grapes and the wines they produce are the product of an unseen microbial terroir, shaped by the climate and geography of the region, vineyard, and even individual vine. DNA tests revealed patterns in the yeast and showed that vineyard environmental conditions can influence bacterial communities on the surface of wine grapes. These conditions include the vineyard location, soil type, and grape variety, as well as flora and fauna in and around the vineyard, like vegetation, flowers, grasses, and even insects. These microflorae release aromatic compounds through enzymatic bioactivity that occurs during the fermentation process. "The study results represent a real paradigm shift in our understanding of grape and wine production and how microbial communities impact the qualities of agricultural products," said Professor David Mills, a microbiologist in the Department of Viticulture and Enology.[2]

Another study published in 2015 by researchers in New Zealand confirmed that local differences in yeast strains can affect the outcome of fermentation. The bottom line is that different yeast strains used to ferment grapes impart specific chemistry to the wine. Studies like these back up what I've claimed for years; native, wild yeasts found in vineyards are unique to that time and place, creating equally exclusive wines.[3]

To make matters even more interesting, there is increasing evidence that epigenetic factors are at work. Epigenetics studies cellular and physiological variations caused by external or environmental factors. In layman's terms, a plant's genetic makeup and subsequent traits can change in response to long-term ecological influences. Researchers in Australia put this theory to the test by comparing wines made from vines of different ages in the same vineyard. They found that wines made from older vines had more prominent fruit aromas and flavors and a deeper color than wines made from younger vines. Subsequent analysis showed differences between RNA molecules in old and young vine fruit cells. One could surmise that the older vines were less susceptible to environmental conditions since they had adapted over time.

The science of epigenetics can potentially become the Holy Grail of terroir. Theoretically, a vine planted on the North Fork will, in time, develop distinct characteristics caused by the specific environmental factors of the vineyard. If—as many researchers believe—this phenomenon affects other attributes of the fruit, such as aromatics, density, and flavor nuance (among others), the effect of terroir may be way more *sui generis* than anyone believed.

For me, terroir is best understood by observing the human condition. Look around the world, and you'll see people of all different shapes, sizes, and colors. Humans belong to a single species and share a common descent. The biological differences between individuals are microscopic—we all are essentially identical twins. Yet depending on where we live, we exhibit differences in culture, nationality, ethnicity, language, customs, and religion. These differences affect beliefs, practices, and behavior and influence our expectations of one another.

In his Pulitzer Prize-winning book *Guns, Germs, and Steel: The Fates of Human Societies*, Jared Diamond states that the growth and strength of culture are highly influenced by the geographical conditions it develops and grows in. His theory is that the human condition results from a chain of developments, including climate, soil, and geomorphology—mountains, rivers, and coastlines—that has slowly helped to form the way we look, where we settle, how we speak, and who we are. Our similarities are biological, but our diversity is the ultimate product of our surrounding environment. This is essentially the human manifestation of terroir.[4]

Humans are also involved in the terroir of wine grapes. People decide what vines to grow and where; we determine factors such as rootstocks, training systems, yield control, soil management, irrigation, and nutrition. All affect the wine in our glass; we are all part of our terroir in more ways than one.

One doesn't have to travel to Burgundy or Bordeaux to see terroir in action. I've tasted it firsthand in many vineyards on the East End and at our Bedell vineyard in Cutchogue, where the exact clone of Merlot, planted in separate areas on our thirty-acre estate on the same rootstock, produces wines distinctly different from each other.

Understanding terroir and how it relates to wine can take time. Being confident enough in one's terroir and what it can produce is challenging and even a bit scary, seeing how much is on the line. But that is precisely what the intrepid winegrowers on the North Fork have done. The work is not yet finished and will continue to evolve, but we have arrived as a region with a distinct style and flair.

It's been part of my life's work to develop a lexicon that describes the flavors and aromas that emerge from our region's wines. I've identified characteristics like seashore, brine, salt grass, wet rocks, minerals, locust blossoms, wet sand, seashell, oak, cherry blossoms, freshly mown grass, seaweed, cedar, beach plums, freshly tilled earth, maple pollen, wet iron, lilacs, honeysuckle, white roses, dried leaves, tar, pine, iron, humus, and fresh hay. It's a list that continues to grow every year. While our wines can express aromas and flavors analogous to ecosystem elements, it's important to remember that terroir only makes its presence known if we allow it to. As a winemaker, my devotional responsibility is to protect and showcase native characteristics on their journey from the earth to the bottle. One can only accomplish this through careful winemaking, which respects the fruit for what it is and does not impose a specific style or production recipe onto it. In essence, one has to let the fruit drive the style. Poor growing techniques, rough handling, microbiological spoilage, oxidation, and too much oak are just a few things that can mask the terroir of wine. Only when we let natural voices sing can the true music of terroir be heard.

Music

Music is the wine which inspires one to new generative processes, and I am Bacchus who presses out this glorious wine for mankind and makes them spiritually drunken.

—Ludwig van Beethoven

The pleasures derived from music and wine have been enjoyed by people for thousands of years. Music has accompanied every vintage I've experienced, from radio, cassette tapes, and CDs to the current incarnation of developing streaming playlists on my phone for specific harvests. Winemaking and music are a perfect pairing; they provide energy and joy during the long and arduous hours spent on the crush pad at all hours of the day and night.

Recently, several studies have shown how music can affect how we perceive the taste of wine. I don't find this surprising, as music can affect our senses in various ways. That's one of the reasons music is so essential to our lives and can provide relaxation, happiness, spirituality, sadness, and even anger. It's part of the neurological phenomenon called synesthesia, defined as the stimulation of one sensory pathway that leads to automatic, involuntary experiences in a second sensory pathway.[1] The sensory equivalent to the old "Dem Bones" song if you will.

Some people even believe playing music can affect yeast and flavor development during fermentation. This premise, I think, is nonsense. But what I have found helpful is our ability to appreciate and understand wine like we do music. I often use this analogy when explaining how to taste wine, especially to beginners. I often say that the choice of what wine you enjoy is quite similar to your choice of music—it is very personal. Music,

58

like wine, can be varied, but some rules exist. Like music, some wine can be just flat-out bad or poorly crafted. There is a baseline for acceptable quality.

Perfumers have long understood the relationship between wine and music, using musical metaphors and referring to bass and treble notes when describing a fragrance. As an amateur musician, I use this philosophy during my winemaking, specifically during the blending process, where wine aromas and flavors can also be categorized into bass and treble elements.

The idea that sounds and scents are linked in the brain was first suggested in 1862 by the British perfume maker G. W. Septimus Piesse, who indicated there is an octave of odor and used a musical analogy to aid in "composing" perfumes. Piesse wrote, "scents, like sounds, appear to influence the olfactory nerve in certain definite degrees."[2]

Individual scent components are "notes"—"top notes," which dissipate quickly, and "bass notes," which dissipate slowly. Chords are built of multiple notes; in less technical conversations, perfumers speak of harmony, balance, and even rhythm. In the old days, the shelves that stored perfumer's bottles were referred to as the "organ" because of its visual similarity to the well-known instrument of the church.

The scientific connection between aroma and music was verified in 2004 by researchers at the Nathan S. Kline Institute for Psychiatric Research in Orangeburg, NY. They found that our perception of sound, smell, and taste can converge and affect us positively and negatively. Simply put, information received through the nose and the mouth is altered by noise. They coined this phenomenon as "smound," the sensory experience created from merging scents and sounds in the brain.[3]

More recently, scientists at the Crossmodal Research Laboratory at the University of Oxford explored how all the five senses interact. Their research showed that implicit associations between taste and pitch might exist. They found that people naturally associate certain aromas and flavors with certain musical sounds, and that taste may be altered depending on the accompanying soundtrack. High-pitched sounds were associated with sweet- and sour-tasting foods, while low-pitched notes were paired with bitter and umami tastes.[4]

The bottom line is that our senses are not as separate as we used to believe; these studies show that what we taste and smell can be heavily influenced by what we hear.

As a winemaker, I find the analogy of using music to help define my craft to be beneficial. This becomes especially true during the process of blending. After harvest, the wines that inhabit the wine cellar express many different characteristics, even within the same variety. The art of blending recognizes all the musical notes in your cellar to determine which can work best together to form harmonious chords and, perhaps ultimately, a beautiful symphony of flavor. Sometimes, there is enough of a melody in just one variety to sound perfect; some wines may express harmony in a small ensemble, while others deserve a part in a larger orchestra.

The North Fork terroir is alive with an abundance of music. There is a symphony of sound inherent in the land ranging from the bass notes of earth and cedar in our red wines to the bright and crisp treble notes of saline, citrus, and flowers in our whites. Many of these aromas and flavors have been described before, but I thought it was time to develop our own notation.

For the first time ever, I present the Musical Scale of North Fork Terroir.

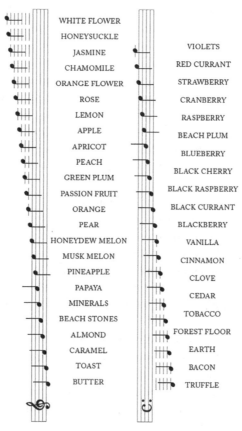

Musical Scale of North Fork Wines. Courtesy of the author.

Minerality

Now I see the secret of making the best person: it is to grow in the open air and to eat and sleep with the earth.

—Walt Whitman

The discussion about where flavors and aromas of terroir originate is a long and winding road. Nowhere has this topic been more debated than when it comes to minerality—the perception of soil or geologic elements in the glass.

For years, a school of thought was convinced that minerality was derived from the vine absorbing geologic characteristics from the earth. The theory was that while vine roots were pulling up water, minerals were also being carried into the grapes, eventually making their way into the wine. This theory remains of prime importance to many winemakers worldwide who discuss soil on their labels and marketing materials to direct the taster to understand what they are perceiving and highlight the uniqueness and superiority of their products. The words people use to express minerality span the natural world. Descriptions include earth, flint, chalk, wet stone, iron, and seashore-related things such as iodine, saltiness, and seashells. The implication is that these minerals provide a littoral expression of place. Today, most wine professionals accept the scientific consensus that minerality is not simply due to vines sucking up geological minerals from the soil and transmitting them to the finished wine for us to taste. Minerals must dissolve in water before a root system can absorb them. These are then taken up as elemental ions in tiny quantities. Nevertheless, the concept of minerality remains a highly fashionable wine descriptor. What, then is everyone tasting?

It's been demonstrated that rocks themselves have no taste. The aroma we perceive from stones is due to bacteria, algae, molds, and lipids that can coat the surface of rocks, producing highly aromatic vapors, especially when warmed on a sunny day or wetted in a shower of rain. Take a dry rock and lick it. You'd be hard-pressed to get any flavor. Individual minerals, however, are another story.

The Monell Chemical Senses Center in Philadelphia has studied the phenomenon of mineral taste. Dr. Michael G. Tordoff, a psychobiologist at the center, stated that "people can taste minerals with biological significance, for sure—for example, sodium, calcium, magnesium, and zinc. Nobody knows how or whether there is some kind of general mechanism that detects 'mineraliness.'" Their experiments have shown that different minerals have unique flavors and can be perceived in extremely low concentrations. Dr. Tordoff also noted that calcium was one of the more perceived mineral flavors. "Some people I have tested can detect calcium dissolved in water in micromolar concentrations, much lower than the levels in wine," said Tordoff. "It wouldn't surprise me if the calcium flavor modified other wine flavors."[1]

The same holds true for metals such as iron. It's been documented that metals do not volatilize; hence they don't have any odor. The smells we perceive from metals only occur after we handle them; metals that come into contact with our skin create a reaction that we can smell from our hands. In essence, we are smelling ourselves.

So, what exactly is going on concerning minerality in wine? Is this all just something associated with our collective enological imagination? Is it just a metaphorical experience? Any good therapist would tell you that your feelings and experiences are real and coming from somewhere. So, where are these aromas and flavors coming from?

I believe the answer is far more prosaic. To understand minerality in wine, one must dive deeper into another common mineral descriptor—salinity.

The concept of tasting salinity, or saline minerality as it's often called, has taken hold over the past decade. Wine journalists have used the term to describe North Fork wines for years. I've even heard it from people who aren't wine drinkers and didn't know anything about the local wine they were tasting. Many people believe that it comes from vines taking up salt water from the ground, but vine roots don't go deep enough to reach salt water, and even if they did, the roots would avoid it as salt water is toxic to vine growth. What we do know is that this phenomenon is present in other wine-producing areas that exist near the sea.

One of the most famous seaside wine regions is northwestern Spain's Rías Baixas, where some vineyards are a stone's throw from the sea. Roberto Taibo, the winemaker of Moraima, in the coastal area of Val do Salnés, stated, "The wind and rain coming in from the Atlantic is rich in chloride and other salts of the sea. This influences the soil by adding chlorides, phosphates, and sulfates, as well as potassium, sodium, calcium, and other salts, which leads to elevated minerality."[2]

These salts are taken up by vine roots in small concentrations. According to a study done in 2019 by the *Australian Journal of Grape and Wine Research*, the recognition threshold for salty taste may lie between a chloride concentration in wine between 300 and 500 milligrams per liter. Wines in Rías Baixas can reach salinity concentrations of 200 to 400 milligrams per liter, which may be what people are tasting.

A 2010 study published in the *American Journal of Enology & Viticulture* tested vines grown on sites with soils of differing salt content. The study found that primary and secondary fermentation aromas are influenced by soil salinity and that increasing soil salinity enhanced not only the saline character but also color intensity, polyphenols, and anthocyanins. Tasters from this study also preferred the wines grown on the more saline sites, finding the wine from the least saline sites "flat and dull." An interesting fact regarding this phenomenon is that lighter soils with higher levels of sand are usually found closer to the sea.

Salinity in wine exists in other seaside regions. In Greece, Giannis Paraskevopoulos, the owner and winemaker of Gaia Wines, tested the salinity levels of two Assyrtiko wines—one grown on the island of Santorini and one grown far from the sea on mainland Greece. He explains that the Santorini wines carried an intense saline minerality while the wines from the mainland did not.[3] He's convinced this effect is because Santorini is located "in the Aegean Sea, which is saltier than the Atlantic." Paraskevopoulos also noted that his Santorini vineyard is "constantly blasted by gusty sea winds that get stronger as we approach harvest." Further analysis showed that the Santorini wines showed over a 200 percent increase in sodium concentration.[4]

These findings back up what I've believed for years—a lot of what we taste in terms of minerality derives from ambient conditions in the vineyard. Minerals and salts can make their way into wine in other ways than through the root system. While there's no research on the topic, I believe that small amounts of soil, salt, and rock dust accumulates on the leaves and fruit via wind, rain, and atmospheric condensation. These components are tangible in the glass.

The North Fork is one of the windiest wine-growing districts in the world, with an average speed of over 11 mph. No other domestic wine region even comes close. The wind speeds increase dramatically during

September and October, averaging over 13 mph. Breezes coming off salt water are a persistent part of our terroir, and there's no question that our fruit accumulates salt from our surrounding waters before harvest.[5]

Studies also show that saline concentrations in condensation increased in a humid environment. As the air on the North Fork already contains a heavy saturation of water molecules in the air, the aroma of the sea is omnipresent. When I've traveled away from the North Fork for any length of time, one of the first things I notice when arriving home is the smell of the sea. These odors are ever-present and not as noticeable daily as we get used to them. Vines grow in this environment every day, and during harvest, the fruit and leaves accumulate salt—both from condensation and from winds blowing off the water. It's also possible that these elements can combine with alcohol, fatty acids, and oxygen during wine fermentation and may be the catalyst for aromas evident in the finished wine.

Along with aerial accumulation, salinity is produced by succinic acid, a weak acid found in wines in the range of 200–700 mg/L, enough to influence taste. In wines with abundant saline characteristics and acidity, salivary glands are activated in the corners of the cheeks, similar to lemon juice on the tongue. Succinic acid is not found in fresh grapes but is an abundant by-product of fermentation (after ethanol, CO_2, and glycerol). Malic acid (which is commonly found in apples but also exists in grapes) is partially converted into succinic acid, and indigenous yeast fermentations have been shown to yield higher succinic acid levels, especially in turbid and warm musts. Yeast species and strains also significantly impact the amount produced; another example of how native yeasts can influence terroir.

Vineyards can be dusty, especially during dry periods where any tractor activity or driving near the vines can result in clouds of soil and plant material. As grapes start to ripen during *veraison*, they begin to absorb aromatics from their surroundings. This becomes more important as the season moves towards harvest, and equipment travel throughout the vineyard increases along with the wind. It's not unusual to see the dust on ripe fruit at harvest, and since no winemaker in their right mind would ever wash grapes before fermentation, microscopic soil particles are present in freshly pressed juice. The combination of yeast, bacteria, alcohol, and lipids present during fermentation produces aromas and flavors from these particles before settling out in the lees. This theory explains the presence of distinct soil characteristics found in wine from different locations and could make up a significant part of what we perceive as terroir.

Nowhere has this aerial phenomenon been seen and studied more than on the West Coast and Australia, where the issue of smoke taint from wildfires has seriously affected wine quality over the last decade. Smoke impact in wine was first identified as a serious problem after the 2003 wildfires in Australia and British Columbia. It became an issue on the minds of California winemakers following the fires of 2008. With the disastrous fires of 2020 and 2021, smoke taint in wines became the top research priority for the West Coast wine industry and remains a major concern for winegrowers in the region.

Smoke aromas and flavors from these wildfires have negatively impacted the quality of wines produced in these areas. Growers and wineries have been affected by smoke originating hundreds of miles away from their vineyards. Burning wood and brush release aromatic compounds called volatile phenols. Once smoke is in a vineyard, these volatile compounds can get onto grapevines and leaves and, depending upon the phase of the growing season, the berries. Grape berries act like a sponge, and these compounds can permeate the grape skins and rapidly bond with the sugars inside to form molecules called glycosides. Freshly picked grapes often don't exude smokiness; however, once the grapes are fermented, the acidity in the resulting wine will begin to break these bonds, releasing volatile phenols. As smoke is detected in extremely low quantities (nanograms per liter), these aromas can be found in finished wines.

Another aerial issue that concerns California winegrowers is the increased growth of hemp production. While research on this topic is still ongoing, many concerns exist in regions with large cannabis-growing facilities. Recent data from the University of Adelaide in Australia showed it is possible that terpenes—the aroma compounds responsible for cannabis odor—could be absorbed into grapes. It has already been demonstrated that terpenes from eucalyptus and lavender planted near vineyards can affect grapes, so the concern over hemp is valid. Anita Oberholster, cooperative extension specialist in enology at University of California, Davis stated that "scientifically, there may be a potential impact." Oberholster said that hemp terpenes could very well "change the character of the wine significantly. Luckily for winegrowers, the aroma from hemp outdoors does not last long and may be greatly minimized by distance. "I think cannabis and vineyards can coexist," Oberholster said. "All we need to do is find the parameters." One thing researchers agree on is that repeated exposure to these types of environmental factors will affect both the fruit and the wine in the glass.[6]

The concept of terroir has been the driving force in identifying wine styles around the world. The fact that something so ubiquitous can still generate skepticism among wine professionals tells me that most people are overthinking the mechanisms at work. The atmosphere of any vineyard is filled with the indigenous, natural aromas surrounding it. Grapes hang on the vines for months and act as sponges, constantly absorbing microdoses of organic materials, saltwater spray, soil particles, aromatic molecules, and microbes. It should be no surprise that these elements can combine to influence the finished wine.

Tasting terroir is all about subtleties. Natural elements are in small quantities and are perceived only by paying close attention. These compounds and their smells are not metaphorical; they are measurable and real in wine and in their natural world equivalents. Of course, the human factor needs to be recognized. Making terroir wines is all about respecting the essence of a place. Intense intervention and manipulation in the wine cellar can easily mask these natural components, suppressing what Mother Nature has provided.

While the North Fork built its reputation by growing the famous wine grapes of Europe, we no longer need to mimic the winemaking practices of other places or manipulate our fruit into becoming a wine it wasn't meant to be. Having been through over forty vintages, I've learned what is possible

Minerality, Peconic Bay Beach. Courtesy of Rory MacNish Photography.

and what is folly. As I've mentioned, no one has made more mistakes making wine than I have. The key is to learn from them and move forward. In the famous words of Albert Einstein "anyone who has never made a mistake has never tried anything new."

Few areas in the country offer the rare mix of sea breeze and freshly tilled earth. These and other native traits, along with our cool maritime climate, allow us to craft elegantly balanced wines found nowhere else in the world. The truth is, we have all we need right here.

American Viticultural Areas (AVAs)

A person's a person, no matter how small.

—Dr. Seuss

According to the official definition provided by the Federal Tax and Trade Bureau (TTB), an American Viticultural Area (or AVA for short) is "a delimited grape-growing region with specific geographic or climatic features that distinguish it from the surrounding regions and affect how grapes are grown. Using an AVA designation on a wine label allows vintners to describe more accurately the origin of their wines to consumers and helps consumers identify wines they may purchase."[1]

Unlike Old World appellations of origin, an American AVA doesn't contribute any parameters concerning quality; it only provides a guarantee of where the fruit was grown. AVA-labeled bottles are also required to contain no less than 85 percent of the volume of the wine derived from grapes grown in the designated viticultural area. I'd prefer a 100 percent requirement, but the Feds always provide some wiggle room in legislation.

As the author of three American Viticultural Areas, I've come to appreciate the rigorous work that goes into an application. It's become a lot easier to gather information in the internet age. Still, one cannot get approval for an AVA unless there is compelling data that shows something distinct and unique about the growing region. To accomplish this, I've taken several deep dives into our region's climate, soil, and history. Much of what I've learned during this process led to the creation of this book.

It was on a Saturday morning in the late summer of 1983 when my boss at the time, Lyle Greenfield, flagged me down in the vineyard while I was driving a tractor.

"Go get washed up; you're coming with me to a meeting," he said. I asked about the meeting in the car on the way there. "It's the local grape growers association, and they want to talk about setting up an American Viticultural Area, an AVA," he said.

"Why do you need me there?" I asked. Lyle was the creative force behind a Manhattan advertising company. He knew a lot about marketing but not that much about how to grow grapes and make wine. "I need your brain," he said.

When we got to the Cooperative Extension office in Riverhead, the meeting room was filled with a few dozen people, men and women, all participants in the newest business on the East End—growing European wine grapes. This was one of the early meetings of the nascent Long Island Grape Growers Association. Lyle gave an impassioned speech about why we should apply for an overall "Long Island" AVA. It fell on deaf ears as the rest of the room all wanted to pursue an AVA geared exclusively for the North Fork. Being the only vineyard located on the South Fork at the time, we were outnumbered. On the way home in the car, Lyle was furious.

"You know what?" he said. "We're going to do our own thing. If I gave you the time and some money for resources, do you think you could write this up?" I was more than ready. I investigated what I needed for the petition and started working on the project during the winter of 1984.

Doing this kind of research in 1984 meant traveling to university libraries and local weather stations for information. I spent hours on the phone speaking with weather station operators and making photocopies of data. One of my most important contacts in this regard was Mr. Richard Hendrickson.

A retired farmer who lived until age 103, Mr. Hendrickson was the nation's longest-serving volunteer weather watcher. He recorded the East End's climatic history from when he was seventeen years old in 1930 until September 2015, when he grudgingly retired at age 103. The information he provided me was instrumental in approving the Hamptons's AVA application and, later on, the North Fork of Long Island AVA.

A Federal AVA petition requires the author to identify a wine region's unique characteristics, including climate and soil data. It also requires a background history of the area, the use of the name, and a physical description of the area. First, you must prove that the name submitted for the AVA is recognized as an actual location—not just made up—then you must prove its existence by showing historical usage. In addition, the grape-growing history of the proposed name must be documented—acres of vines under cultivation, what the future might look like, economic impacts, etc. This wasn't a problem at all. Interestingly, I found information on grape growing from some nurseries in western Long Island dating back to the early 1700s.

Then, of course, a case needed to be made that the AVA is geologically and climatically unique from adjacent areas. This is the meat of the application and is done using local soil surveys and weather data. It is time-consuming and takes a great deal of research and writing. The information provided by Mr. Hendrickson and the Cornell Agricultural Research Station in Riverhead made it possible to write up an argument that the South Fork had growing conditions distinct from the North Fork and led to the approval of the Hamptons, Long Island AVA in 1984.

Ironically, while the Hamptons AVA sailed through the approval process, the application submitted by North Fork languished. After being rejected multiple times for being incomplete, the grape growers association asked me to write up the North Fork AVA. When I asked for a copy of what they submitted, I was given a one-page letter addressed to the BATF, stating that the North Fork needed its own AVA. I knew what had to be done as I had already accumulated all the data required, so it wasn't as tedious to complete as the Hamptons application. The North Fork of Long Island AVA was approved the following year, much to the relief of local growers.

Years later, I felt it was time to revisit the original idea put forward by Lyle Greenfield and began work on a more extensive Long Island application. There were several reasons to do this. Grape growing areas without an AVA are limited to using either their county or state name, neither of which was an attractive option. An overall Long Island AVA gave wineries in both East End AVAs the ability to label their wines appropriately if they purchased fruit from each other. It also provided a growing number of vintners located west of the two East End AVAs a way to label their wines. Most importantly, it helped protect the *Long Island* name from being misused. Without an official AVA, the name *Long Island* remained unprotected. It could theoretically be used by companies that wanted to sell wine in our local market that wasn't grown in our region.

So, with the blessing of Raphael owner Jack Petrocelli, I started researching the requirements for a Long Island designation in late 1999. I didn't have to go to the library this time as all the data was available online. Unsurprisingly, a few growers weren't happy I had pursued the application, and I received little support from the Long Island Wine Council. The application was submitted and approved in the late spring of 2001.

From that point forward, consumers could be assured that any wine named *Long Island* would indeed be made from grapes grown in the region. With the reputation of Long Island wines improving every year and demand rising, it seemed like an easy decision to me.

With the advent of satellite-based weather recording, there's the possibility for additional, smaller AVAs in the future. The western part of the North Fork has a different dominant soil type (Riverhead Sandy Loam) than the eastern part, which is composed mainly of Haven Loam. Temperatures are slightly higher near Riverhead and get slightly cooler towards Orient Point, so there is a slight difference between these three sections of the Fork. However, more data is needed to make a case for these smaller and separate AVAs. It's part of the natural evolution of a wine region. As more knowledge is gained about the site and soil, there could be a movement to delineate the North Fork further and drill deeper into the distinct weather conditions at either end of the peninsula.

American Viticultural Areas are the only federal designations given to the domestic wine industry. While they don't guarantee any level of quality, they provide legal definitions of the region, protecting the geographical boundaries of where the grapes for a wine are grown. In addition to providing consumer protection, they help define a broad area of terroir as the approval of an AVA is based on the distinct climatic and geologic conditions found in the region. Knowing the style of wines made from different AVAs brings more information to the consumer. For states with significant wine industries like California and New York, a state-only name on the label provides little guidance on what kind of wine is in the bottle. An AVA is the first step in helping consumers understand what terroir is. Of course, the last step in this process is tasting the difference. Please see the appendix for more detailed information on the three Long Island AVAs.

Winemaking

Wine is constant proof that God loves us and wants to see us happy.

—Benjamin Franklin

When I first meet people and tell them I'm a winemaker, they're usually surprised. It's not a profession people run into every day, especially on Long Island. I'll get questions like, "Are your feet always purple?" or "What do you do in the winter when all the grapes are picked?" and my favorite—"It must be nice to have a job where you drink all day!" I explain that none of that happens, cellar work keeps me busy twelve months out of year, and I spit out wine all day. I do tread on grapes every once in a while, but with boots on. Our wine culture in America is still comparatively young, and the profession has only existed on Long Island for less than a generation. Naturally, people have many questions about what we do for a living, and most have an overly romanticized idea of the process.

A woman who worked in our tasting room once expressed interest in becoming a winemaker but was doubtful she'd be able to achieve her goal since she didn't have any international experience. "Everyone travels to all these wine regions to work harvests before working as winemakers and sommeliers," she said. I assured her that becoming a winemaker doesn't require global work experience. The recent fad of working harvests in famous wine-producing regions has more to do with travel and lifestyle than developing a fundamental understanding of the craft. Winemaking requires a basic knowledge of biology, chemistry, and the agricultural sciences. It's farming on an intense and grand scale. Don't get me wrong, it's great to see other regions and learn from them—but make no mistake—learning how to make great wine is all about what's going on right under your feet. The best winemakers in the world learned their trade by studying the climate and soils of their regions and understanding the conditions where they live and work. That kind of inquiry is hard to do when you're moving around the globe.

Most importantly, being a winemaker is nothing like being a sommelier. Understanding many different wines from around the world is an excellent skill to acquire, but it really can't help you become a great winemaker. A sommelier is more like a librarian or an art curator, while a winemaker is more like an author or a painter, combining their ideas with the world around them and making them a reality. One requires a broad understanding of the subject, while the other is about the specialization of a craft. One is evaluative and organizational, while the other is agricultural, scientific, and creative. Both are important jobs but completely different from each other. Being a winemaker means figuring out every detail about where you want to make wine. That's how great winemakers are made—not by trying to guess what wine is in front of you like some kind of game. But a great winemaker needs to know what great wine tastes like to understand the potential of one's own enological landscape. It's about specializing—drilling deep into a place and understanding how to bring out the best in the fruit.

The best winemakers in the world don't have much time for travel. They probably know less about the wines of other great regions than your average sommelier. Instead, they have honed their skills in their own vineyards. In the case of the Old World, many local techniques come from time-honored traditions that have been in place for generations. The bottom line is this—great winemakers are made at home.

My path to winemaking started literally from the ground up. Coming out of Cornell, my academic training was geared toward the agricultural aspects of growing wine in the vineyard. In the East, it's critical for winemakers to be well versed in the vineyard in order to make fine wine in the cellar. We have an Old World model—a vigneron rather than a winemaker. The typical university model of training winemakers in the US provides little real instruction in viticulture. Many West Coast–trained winemakers I've met feel they don't need to truly understand viticulture as these warmer regions have more predictable weather patterns with minimal vintage variation. Unlike other New World regions, the temperate North Fork has four distinct seasons with cold, heat, rain, snow, humidity, and lots of disease pressure. Thankfully, we also have steady winds and well-drained soils that allow the vine roots to stay relatively dry.

Making wine in temperate regions with humidity and seasonal rainfall is nothing new; European vignerons have been doing it for centuries. But growing European grapes in a cool, North American maritime climate is much more challenging, making it all the more rewarding.

Winemaking on Long Island is not for the faint of heart; more than one California winemaker working on the North Fork has left with their tail between their legs. Our weather can be unpredictable and sometimes unforgiving, but as winemakers in the *Cool* New World, we must be flexible and react accordingly. I've often said North Fork winemaking is akin to playing jazz; every time you play a song (or experience a grape harvest), it's slightly different from the time before. Reading music doesn't help when you're in the middle of an improv performance.

When I first started making wine in Bridgehampton, I was also in charge of managing the vineyard. During my first few harvests, I found mostly local people, old and young, to help pick grapes. Most had never even seen grapes growing on the vine before, so they needed to be shown what to do. Grape stems are tough and need to be cut with a knife or clippers, and we harvested everything by hand into small buckets and stackable baskets. We'd pick all day until the sun started to go down. As soon as we finished, I'd jump on the tractor and drive down the rows to pick up the baskets. Once the fruit was unloaded at the winery, I'd empty the baskets into a crusher/de-stemmer and then into a one-ton Howard Roto-Press. It was a medieval-looking machine made of orange-painted steel with circular iron plates connected by chains and powered by incredibly wonky electronics. Whenever this machine broke down, which was frequently, it was always an adventure watching our local electrician trying to figure out the German circuit breakers and wiring configurations. If all went well, I could move the freshly pressed juice into a stainless tank, hoping it would be cool enough to settle out until the morning. Since I had no refrigeration, I wrapped canvas soaker hoses around the outsides of the tanks and let cold well water slowly dribble down the sides until morning. More often than not, the juice began fermenting on its own, presenting me with an early introduction to the world of native fermentations.

After washing all the baskets and equipment and shoveling up all the pomace into a manure spreader, I'd jump back on the tractor to spread it over the field with only the light of the moon to guide my way. Once finished, I'd drag myself home, leaving my fruit-fly-infested clothes outside while I ate, showered, and slept, only to get up early the next morning to do it all over again. That is what being a winemaker is all about. I was living my dream as a vintner and loving every minute of it.

Winemaking requires working long days, through heat and cold, with soggy, wet feet, juice-covered pants, and painful hands. When I'm in the middle of harvest and go out in public, more than one person has seen my hands and questioned if I was a car mechanic. Making wine involves hauling tons of fruit and pushing your body until all hours of the day and night until you go home and collapse—only to have to get up early the next day to do it all over again—for weeks and months on end. It's about sacrificing relationships with friends and family during the fall while being at the complete mercy of the weather, understanding that no matter what type of plan you had, Mother Nature usually changes it.

My time at Bridgehampton was a real learning experience. I learned how to make wine with minimal equipment, materials, and labor. I was involved in every aspect of production and understood how much work needed to be done to create a world-class wine. I also did a lot of experimentation with techniques, grape varieties, yeasts, oak, fermentation temperatures, and variables. I made some mediocre wines but, in the process, learned what it takes to make world-class wines and grew to understand our terroir. As I began to work for companies with greater resources, I implemented much of what I learned along the way. But I never forgot where I came from, and I always knew I could make do with much less if I had to.

The person I learned the most from was the late Paul Pontallier, the managing director at Château Margaux in Bordeaux. Paul was an expert viticulturist and enologist who understood the complex relationship between place and wine. His interest in how grapes respond to environmental and geographical conditions led him to join Raphael as a consultant.

A student of the world, Paul was interested in educating himself as he helped me. Pontallier was also mindful of grapes' resilience and ability to make fine wine. At one of our first meetings, where I rattled off statistics about ripeness levels, pH, acidity, and other numeric indicators that often infiltrate a winemaker's lexicon, Paul listened and quietly said, "American winemakers get too hung up on numbers. You should start by throwing away your pH meter!" I had just bought a brand-new meter and couldn't bear to throw it in the trash, but his comments helped me to pay more attention to the taste of the fruit and wine and less time looking at the lab analysis. Paul understood that great wines are not made with a recipe. "If the best wine you can make means keeping it in stainless steel and not oak, then that's what you do," he insisted. It was music to my ears.

There aren't many new ideas in the wine world. Even the most avant-garde winemaker utilizes techniques that have been done before. Many ancient practices, like *pétillant naturel*, orange wine, and piquette, have become trendy in winemaking circles over the last few years. These practices sometimes lead to more interesting wines from prosaic grape varieties. Most often, we're translating old techniques to fit a modern aesthetic. We are learning why certain techniques lead to particular outcomes through trial and error and scientific analysis, a pathway that has guided winemaking for centuries. Through both experience and experimentation, anecdotal evidence can become scientific facts. This approach is a part of sustainability and guides winemaking and other aspects of our lives to discern the best way forward.

For each vintage, I approach the winemaking process with an open mind. I allow the history of the fruit I'm working with—the specific block or vineyard—and the vintage season to take an informed approach to wine style. For example, I won't try to make a big, extracted wine out of fruit that doesn't have that potential. In all cases, I don't alter the natural condition of the fruit and its ripeness level. If, as it sometimes happens on the North Fork, our fruit can ripen to only 19 or 20 percent sugar (also known as degrees Brix), I leave the fruit alone and acknowledge that the wines will be lighter than they would be at higher ripeness levels. I trust that the wine will be balanced and won't add more sugar, which would increase the alcohol content and create disharmony in the bottle. Adding acid is another nonstarter for me. Most of the time, it's best to trust the terroir and leave well enough alone. The North Fork conditions provide everything we need to make this philosophy a successful reality.

For me, it's always been more about style than variety. North Fork wines have a bright, delicate fruitiness, which can be easily overwhelmed by an excess of new oak barrels, making older, neutral barrels important. Oak treatment should be done judiciously and be completely integrated with the wine to provide an unassuming, nuanced structure. New oak is typically not our friend and needs to be used carefully and sparingly. I want my fruit to sing, and North Fork fruit has a delicate, ethereal voice; new oak is like a loud backup band. If I can taste it outright, it's a flaw to me. I love using mostly older, neutral barrels, which allow extremely restrained and slow oak character into wine while still providing the much-needed slow oxidation to round out the tannins and soften the mouthfeel. When I purchase new barrels, I go for the tightest grain possible and the least amount of flavor extraction over time. Most of the time, these barrels are French; however, plenty of French barrels are overwhelming for our wines. Rich, ripe reds can benefit from some new wood except for Cabernet Franc, which, in my opinion, never improves in a new oak vessel. Utilizing and properly maintaining older barrels is, therefore, essential.

Many winemakers talk about intervention. Some use the term as a pejorative, as if wine could make itself without our input. Making wine is an interventionist activity, and wine would not exist without our help. A winemaker's careful direction helps guide the fruit into the bottle.

Probably the most crucial decision a winemaker can make is when to harvest. Unlike many other fruits, grapes don't continue to ripen after being picked. Goals for harvesting can be important, but it's even more critical to know when your fruit has ripened completely before it starts to break down. Hangtime on the vine is important; however, there is a point of decreasing returns. Hand-harvesting remains the best option, as a critical eye makes better decisions than a mechanical harvester. Also, as varieties within the same block can mature at different times, handpicking allows for more flexibility. Some winemakers even pick one side of a vine row, lifting the bird netting on one side to harvest, leaving the other half to ripen fully. Machine harvesting doesn't allow any of this. Thankfully, nighttime picking is something that never needs to happen on the North Fork, as daytime temperatures are typically cool enough for harvesting. Other decisions follow, such as whether to crush and destem or press the whole clusters, how hard to press, what temperature to ferment, how often to pump over, and when to bottle. A vigneron trying to make terroir wines will always let the fruit lead this process.

Another example of this type of intervention is managing skin contact time and barrel-aging regimens. I have a few vineyard blocks of Merlot and Cabernet Franc that consistently make a better rosé than red wine. Information is gathered through the experience of working with a particular vineyard. Often, there might be sections of our Chardonnay plantings that, once fermented, do not live up to my expectations for a varietal wine. In cases like this, I'll use the wine in our entry-level white blend instead. With a proper lineup of wines, every fermentation has a home. I approach vintages as I would a labyrinth. You don't fight against it. Instead, you go where it takes you. Invariably, the goal is to make a wine with great flavor, balance, and proper structure. Ultimately, I want the wine to be delicious and have people return for more.

We're blessed to be in a place that has one of the mildest climates and gentlest weather patterns in the Northeast. That, along with some of the finest soils in the world available to grow grapes, allows the region to produce a wide range of European varieties successfully. The potential for diversity is one of our strengths; it's hard to determine the "best" grape for the region as we grow so many so well. Unlike in the Old World, there are no restrictions on what we can and can't grow. Other domestic cool-climate regions have a limited number of *vinifera* that can survive; in those areas, it's a bit easier to get behind a signature grape.

Although the North Fork is one of the few regions on the East Coast that can produce world-class red wines, we have a lot of successful white varieties, and the list is growing. Early on, many of us thought Merlot might be our signature variety; however, in the past few years, with new clonal/rootstock selections, grapes like Cabernet Franc, Malbec, and Petit Verdot have shone. One could make a case that Petit Verdot can make the most exciting and robust red wines here, and as our climate warms, our ability to ripen later-ripening varieties will no doubt expand. Overall, our strength as a region lies in the French-Atlantic varieties that have grown so well in our district. Sauvignon Blanc, Albariño and Melon de Bourgogne make some of the region's most vibrant and compelling whites.

The wines I make are grown in a cool maritime climate on the East Coast of the US. That's a unique situation and a huge accomplishment. To make delicious wines, we don't have to add sugar, acid, or water to our grapes. That's not something many New World wine regions can claim. I want my wines to be a sincere reflection of the North Fork and showcase the low alcohol, crisp, natural acidity, and elegant aromatics that can only be found here. No matter the variety, the essence of the North Fork always shines through—finesse, elegance, balance, salinity, and refreshing acidity. I want my wines to pair well with food and have little to no noticeable oak flavor.

They're a direct reflection of the people who live on the East Coast—slightly rustic but elegant, approachable, and friendly yet somewhat edgy. Frankly, I couldn't imagine making wine anywhere else.

Being a winemaker means you only have one chance a year to create something special—and if you don't—you have to wait a whole year for another opportunity. No matter how much you learn and how much science you master, there will be variables you will not be able to affect. It's best to accept what nature has given you and shepherd the fruit into wine the best possible way you can. And after all that, when you feel you've done the best job you can do—it's about understanding and accepting that no matter what you do and how hard you work, someone, somewhere, may not particularly like the results. You must have the courage of your convictions and enough belief in your craft to know that your creation is beautiful and delicious. Luckily, if you've done your job well, most people will agree with you. Winemaking, as it turns out—is all about the famous prayer: "Grant me the serenity to accept the things I cannot change, courage to change the things I can, and the wisdom to know the difference."

Balance and Style

So divinely is the world organized that every one of us, in our place and time, is in balance with everything else.

—Johann Wolfgang von Goethe

Once, for my birthday, my son gave me an unusual wine holder that I keep on my desk to this day. It's a piece of chain that rises vertically and is made to hold one bottle of wine. On its own, the chain just lays flat, but when a bottle is placed in the opening, the bottle magically holds itself up in perfect balance. I see this bottle daily and use it as a metaphor for my work in the cellar. Because what we do on the North Fork has everything to do with balance. It's what I believe to be our inherent strong point on the North Fork and a wine style that we produce naturally.

Balance regarding wine means a harmonious blend of components: acid, sugar, tannin, and alcohol. For a wine to be considered balanced, all these parts must exist in the proper ratio to each other. Each element plays a vital role in creating a good bottle of wine; too much of one part and the wine will suffer; not enough of a specific component and the wine will be lacking. Both instances are examples of unbalanced wine.

A winemaker can achieve balance in one of two ways; manipulate components artificially or produce wine with the elements found naturally occurring in the fruit. I prefer the latter. Hundreds of additives are available to today's winemakers to affect wine flavor, aroma, and balance. The oldest additive of all is sugar, which has been used for thousands of years to increase alcohol levels in underripe fruit. Products like tannin and powdered acid additions are commonly used to tweak mouthfeel and flavor. When fruit is harvested without some level of balance, there is something wrong that needs to be corrected. Generally, well-grown North Fork grapes need no such corrections.

The recent appreciation of balanced wine illustrates a movement becoming integral to the tastes of the American consumer. Wine drinkers are becoming tired of high alcohol, overly extracted wines. Drinking and tasting high-alcohol wines can lead to palate fatigue, numbing taste buds and making it harder to determine specific flavors. There are also health issues and caloric concerns with higher alcohol wines.

In response, California winemakers are scrambling to produce wines with more finesse and lower levels of alcohol. The problem, of course, is that the climate of California can't easily create this style of wine, and it's getting even more difficult as our environment continues to warm. One can do it—but not naturally.

The West Coast push to produce lower-alcohol wine started in the late 1990s as the region's temperatures started to rise. It began with spinning cone technology and reverse osmosis, eventually evolving into the less-expensive practice of adding water to fermenting grape juice. The amount of viticultural work being done in this field clearly shows that producers are aware of climate change while sensing a shift in the wine marketplace. The trend in California to produce wines with less alcohol and more acid additions is a calculated manipulation of juice and wine to meet new market expectations. These are not terroir wines.

There's no question that a lower-alcohol, refreshing style of wine has gained favor in the marketplace. This perception became clear in an article I read a few years ago in the *New York Times* about local wine in restaurants. The article interviewed a San Francisco wine director who explained his preference for European wines over the local choices found in nearby Napa. He said, "At our restaurant, you need low-alcohol, high-acid wines, and they don't come from the New World." I guess he's never visited the North Fork. *Wine Spectator* also made the case that America's wine palate is slowly trending toward lower alcohol, crisp and elegant wines. These types of wines are right in the North Fork's wheelhouse.

I've been making these kinds of wines for a long time. At Bridgehampton, I produced one of the East Coast's first stainless steel Chardonnay in 1982. It was delicious, but I don't think the market was ready for it. During those days, consumers wanted rich, oaky, and sweet Chardonnays, and many people found my version lacking. "Where's the oak?" would always be the comment. I eventually had to learn to become comfortable in my own skin and make the wines that suited our climate, emphasizing the balance between alcohol, acid, aromatics, and elegance.

When compared to our contemporaries in Europe, the East Coast wine industry is one of the most diverse in the entire world, with *labrusca*, *vinifera*, and hybrid varieties all under cultivation. In France, under the Appellation d'Origine Contrôlée regulations, only approved varieties can be grown in certain areas. While European regions have become the model for winemakers to emulate, weather conditions may not always be favorable for these specific varieties to produce their best wine every year, leaving substantial vintage variation.

Unlike these Old World regions, North Fork vintners aren't regulated in this way and can grow whatever grapes they want. Our climate ripens an array of varieties, and because of this diversity, we can produce delicious wines every year. Vintage variation does exist; however, we can showcase the strength of the vintage through our multifarious lineup of wines and through the craft of artful blending. In the years where rainfall and cool weather dominate, white wines rule the day. I find these vintages extremely interesting and exciting because they offer a winemaker the ultimate test—the ability to adapt to changing conditions and guide the wine accordingly to allow it to retain its optimal balance. Hot and dry years on the North Fork will always be the best years for reds. There are also vintages where both reds and whites are of extraordinary quality (i.e., 2010, 2015, and 2019).

Years ago, when I asked my friend Stephen Mudd about what he thought a "normal" North Fork vintage was, he laughed and replied, "Normal is a cycle on your dryer." Every vintage on the North Fork is different, yet we will always produce high-quality wines due to our diverse lineup of grapes. It's a testament to our climate and soil and to the hardworking people in our vineyards and cellars who strive to keep pushing the quality of North Fork wines higher and to whom every year is filled with great expectations.

After almost five decades of trial and error, a regional style has developed for North Fork wines. Flavors and structure derived from the local environment are reflected in the wines we make—no matter what that variety or blend happens to be. Our white and rosé wines are defined by aromatic, mouthwatering flavors, full of zesty, natural acidity with waves of saline minerality. North Fork reds are medium-weight yet energetic and taut, with velvety tannins wrapped around crunchy, spicy fruit. No matter the variety, our wines will be lower in alcohol, less intoxicating, and more refreshing than anything made on the West Coast. These are the kinds of wine we naturally make. Warmer New World areas facing ever-increasing climate challenges will have a tough time producing wines of this style in the future.

New York has always been where the best information, inventions, food, and fashions have come to be tested and accepted. We continue that age-old tradition in our wines and have passed the test. We know this from our great reception in the marketplace and also because the type of wines we make are now being copied by our friends in the West.

With more than forty vintages under my belt, I've been blessed to have lots of chances to get it right. Here are a few other things I've learned:

• Cool-climate fruit can be fragile once vinified. Careful handling in the cellar is key to maintaining ultimate quality. Rough handling at harvest and aggressive pumping during maturation and aging can bruise and degrade delicate flavors.

• Our vintages are like a jazz performance. A technique that worked well the year before doesn't mean it will again. It can be the same tune, but you may need to play it slightly differently.

• Flavor takes time. Like in any fermentation, from beer to bread, the relationship between yeast and flavor increases with time. Long slow fermentation for whites and extended maceration time for reds are keys to releasing flavor and aromatics. Time on the lees for both can add density and mouthfeel. Cool climate ferments can take months to develop fully.

• North Fork whites typically reach peak flavor and aromatics after a year in the bottle. Varieties like Chardonnay and Gewürztraminer have shown incredible ageability. Reds reach their peak at least five years after the vintage and can last years longer.

• Make important winemaking decisions after tasting your wine away from the cellar. Too many winemakers taste at the tank or barrel and make decisions from that experience. Wine, like a child, behaves differently away from home. Since that is where they will spend most of their lives, we must understand what they're about before intervening.

• Yield is critical. I don't care how great the vintage or site is or how well-managed a canopy you have, the optimal yield level for quality on the North Fork is always between 2½–3½ tons per acre. Lower yields than these are unnecessary in a mature vineyard and won't lead to further ripening. Higher yields, however, will always lead to some level of dilution.

• Being a winemaker is very much like being a chef. It doesn't matter how much training or fancy equipment you have; if the raw materials aren't suitable, the finished product won't be either. Likewise, an untrained winemaker can screw up even the best fruit. On the North Fork and other cool climate regions, good vineyard management and good fruit quality are essential for success.

• In cool climates, wine quality is highly influenced by vine clones. We don't have the benefit of excess sun and heat, which can mask clonal differences in warmer climates. Clones and clonal/rootstock combinations are important for every variety. Some of my favorites are Cabernet Franc 214, Merlot 181, Chardonnay 75 and 95, and Malbec 578. I love reds grafted on the *riparia* rootstock as I'm convinced it leads to slightly smaller vines and accelerates ripening—a good thing for reds in our part of the world. I prefer 101–14 rootstock for whites; most whites ripen earlier than reds and generally can carry a slightly larger crop load.

• Our wines can basically make themselves. One just must work mindfully. We don't need to add sugar, acid, or anything else. Organic yeast food during fermentation helps preserve terroir by reducing hydrogen sulfide production. Sulfur added after fermentation helps preserve terroir by inhibiting oxidation and most spoilage organisms.

• While I believe that wild fermentation is one of the best ways for terroir to shine, natural winemaking is not. Eschewing minimal amounts of sulfur may work for a short while, but in the long run, a cellar can become infested with spoilage organisms such as *Brettanomyces* and *Leuconostoc* bacteria, wiping the terroir out of every wine it touches. Sulfur also reduces oxidation and aldehydes, which also mask terroir. It's a thought-provoking idea, but I think my craft is more than just sitting idly by while a perfectly good wine spoils.

• The fruit should lead. If you want to make a big barrel fermented Chardonnay or bold and powerful Merlot, make sure your fruit is up to the task. Letting your intentions lead the way or forcing a particular wine style will often end in disappointment. Fruit harvested with lower sugar, tannin, and the color is best left alone and allowed to be elegant and restrained. If you have good fruit maturation, go for a more powerful style of wine.

• While it's important to be mindful of the materials used in the vineyard, it's equally important to know what your vines need to survive and be healthy enough to ripen high-quality grapes.

• The old adage that "stressed vines produce better wines" is often misconstrued. The better phrase to use is *balanced*. Like any other plant, vines must be healthy and have adequate water and nutrition to produce quality fruit. A vine suffering from water stress or lacking essential nutrients will not be able to ripen fruit properly and will produce inferior wine.

• Understanding one's terroir not only results in more interesting and delicious wines, it also allows a winemaker to really spread their creative wings.

• Being a winemaker on the North Fork requires a thick skin. When I started out, my wines and home region were negatively critiqued more times than I can remember, and it still happens occasionally. I was yelled at by a few wine shop owners, incensed that a local wine sold for twelve dollars a bottle. One needs to have the courage of their convictions.

Native Yeast

Come away, O human child: To the waters and the wild with a fairy, hand in hand . . .

—William Butler Yeats

Indigenous, spontaneous, natural, aboriginal, feral, native, endemic, ambient, or wild—no matter what you want to call it, the art of making wine without adding commercial yeast is gaining favor in the US. Of course, the use of spontaneous fermentation is nothing new to the cellars of the Old World where wine has been made with natural yeast for thousands of years. Many of the world's finest wines are produced with native yeasts. These winemakers see indigenous yeasts as integral to the authenticity of their wines and feel that natural fermentation imparts a distinct regional character. As the first winemaker on Long Island to produce wines exclusively with native yeasts, I not only agree with my friends overseas, but I also believe it is the fundamental secret to terroir.

I was born with indigenous yeast. Actually, we all were. It's on our bodies, in our homes, and in the air we breathe. But while yeasts are ubiquitous, they are also primarily inconspicuous in our daily lives. As a native Long Islander, I never thought I would use these same yeasts to produce local wine—and that these yeasts would be our very own, found nowhere else in the world.

Yeast microbes are one of the earliest organisms domesticated by humans. We've had yeast around us for thousands of years, and as we developed techniques to make bread, beer, and wine, we unknowingly selected the strains that worked best. It wasn't until 1857 when Louis Pasteur proved that alcoholic fermentation was conducted by living yeasts and not by a chemical transformation, that the use of cultured yeasts began to develop. Since then, the business of isolating and producing yeast on a commercial scale has grown dramatically, with wine producers among the biggest customers.

But commercial winemaking by inoculation can be sort of like painting by number. Yes, the pictures can look nice when you're all done—but did you genuinely create anything? The seemingly endless supply of congruent yet prosaic wine led many winemakers like myself to look hard at what they were doing and ask—"am I getting the most out of my fruit and letting it reflect the place it was grown?" The use of indigenous yeast is a way for producers to cut through the din of the modern marketplace and the international style that has gentrified the wine world. It takes you back home to where you belong.

So, what is it about indigenous fermentation that makes it so unique? For me, it is essential in the quest to define terroir because it is just so—a product of its own environment. In essence, terroir begets terroir, or as they say on the North Fork, "you plant potatoes, you get potatoes."

Every vineyard and wine cellar in the world generates its own serendipitous mix of microflora, all of which can produce unique wines for the estate. The levels and ratios of native yeast populations are determined by their environmental conditions, resulting in a highly localized evolution. While *Saccharomyces* is the primary wine yeast found worldwide, many other yeast species are present in vineyards, including *Hanseniaspora/Kloeckera*, *Candida*, *Cryptococcus*, *Torulaspora*, and the ever-unpredictable *Schizosaccharomyces*. Although *Saccharomyces* can live in the vineyard, they are found in larger quantities in winery buildings and equipment. Non-*Saccharomyces* species are found mainly on the fruit and in freshly processed must.

All the factors in and around a vineyard and winery are at play in determining which combinations of yeasts are present at any given location. These include the grape variety, fruit ripeness, rainfall, condensation, and humidity, as well as the vineyard's canopy management and disease-control strategies. In the cellar, the types and placement of equipment, the temperature and humidity of the building, and even the water's pH will affect the yeast species' ecology. Every winemaker has designer yeast at their disposal, right on their fingertips.

With cultured inoculation, indigenous species are either killed off with sulfite additions or overpowered by avaricious commercial strains. During indigenous fermentation, yeast strains such as *Hanseniaspora* and *Candida* dominate early, eventually giving way to any number of *Saccharomyces* strains. Recent studies show that non-*Saccharomyces* yeasts can profoundly influence aromatic esters, higher alcohols, acid metabolism, and glycerin production.[1] With natural fermentation, native species are allowed to thrive, and with so many more ambient organisms at work, the potential exists for an endless profusion of compounds creating an exclusive character that can't be reproduced anywhere else. Wild yeasts also grow more slowly and undergo a certain amount of stress in completing a fermentation. This struggle produces a certain level of complexity.

While I've believed in the influence of indigenous yeast on wine for many years, recent research has lent validity to this claim. While this work is relatively new, many researchers are now studying microbial terroir's effects on grapes and wines. Research done at Cornell University confirms that microbial populations vary according to geographic locations. Other studies from the University of California at Davis demonstrate that grape microflora and the wine metabolites are regionally distinct.[2]

Typically, a cool, maritime climate like that of the North Fork will breed more significant populations of non-*Saccharomyces* than drier and warmer areas. Over the past few years, more than one California winemaker has commented that the increasingly hot temperatures and arid conditions experienced in their vineyards resulted in little to no natural microflora, necessitating the need for commercial inoculation. The North Fork however, as well as the humid East Coast in general—is fertile ground for natural yeast production.

Importantly for my wines, the vineyard at Bedell has its own brood of native yeast. Some live in the winery while others enter from the vineyard. Many yeasts will reunite each year like old friends and family, while others will be saying hello to each other for the first time. They evolved and established themselves—unbeknownst to us—in sustainable proportions to survive and prosper. In an estate winery like Bedell, terroir comes full circle through the ecology of native fermentation and manifests itself in the glass. How could you possibly drink anything more local?

I want my wine to be unique—to reflect the place in which it was born. To me, local flavor is one of life's great delights. I'm proud that no other place—no other person—can produce wines like the ones I make. It is the ultimate confluence of art and science—a natural design where we are just another cog in the wheel. I like to think that I raise my wines like I did my children, and I know of no better way to begin the creative process than by allowing our native yeasts to provide the genesis. Like my children, I let my wines be who they want to be, explore their interests, and make their own friends. Like any good parent, I keep an eye on them in case they get into trouble. That's my job, and most of the time, when raised with the proper care from the beginning, they'll turn out great, with their own strengths and personalities. Just like my kids, my wines are from the North Fork. They are born with indigenous yeast, and they are locals. In the words of another native New Yorker, Chip Taylor: "Wild thing—you make everything groovy."

Native Yeast Culture at Bedell. Courtesy of Steve Carlson, Bedell Cellars.

Natural Wine and Biodynamics

Skeptical scrutiny is the means, in both science and religion, by which deep thoughts can be winnowed from deep nonsense.

—Carl Sagan

I believe the way to make great wine is to do it as naturally as possible. I practiced this philosophy before "natural winemaking" became a thing. Utilizing sustainable and organic growing techniques, allowing native yeasts, and pursuing a path of minimal intervention are all part of seducing the best that nature provides, allowing the terroir to shine. However, the fact remains that winemaking is inherently interventionist; wine cannot make itself out of grapes any more than bread can create itself out of grain. Both are natural products yet only exist through the action of human hands.

The most passionate debate in wine today is centered on the concept of natural winemaking and the various techniques used to create it. Whether termed raw, live, pure, or authentic, disciples of this movement say they produce wines "the way they used to be," without any additives or human manipulation. Some wine writers have called natural wine "not new" and "what wine always was."[1] While I greatly respect the shift away from overly technical wines, the natural wine movement has become increasingly dogmatic, insisting on a black-and-white definition that has divided the industry. More specifically, the crux of the natural and raw wine ideology seems to center around using one additive—sulfur. But the truth surrounding all of this is a lot more complicated, and the past is not as idyllic as some would believe.

The story of ancient winemaking before the discovery of sulfur was one endless quest to preserve wine and keep it from spoiling. No doubt the first wines were made from the spontaneous fermentation of grape juice.

People found that drinking this tart beverage created a pleasing, intoxicating effect, but the drink wasn't enjoyable for very long, eventually turning sour and bitter. Since time immemorial, additives have been used in wine to keep it from oxidizing and turning into vinegar. Early producers used materials like lead, lye-ash, marble dust, salt, tree resin, pitch, fish garum, and even salt water in their wine to try to preserve it. Ancient wines not only had more additives than modern wines, but they most likely tasted terrible based on present-day standards. It wasn't until the advent of sulfur, first documented by the Homeric Greeks and Romans, and eventually fine-tuned by the Germans and French, that the era of fine winemaking began.

Natural wine ideologues would have you believe that scientific progress has been destructive to modern winemaking. They argue that we have moved too far from a past ideal; some even say drinking anything other than "natural wine" will harm you. There's even an online newsletter with the hyperbolic title of "Not Drinking Poison." They want you to believe that oxidized wines full of microbiological flaws are healthier for you and offer the most straightforward pathway to exhibit terroir. I profoundly disagree on all points and so does the science.[2]

The grand irony is that the modern natural wine movement is an artificial construct built on the foundation of scientific improvements in enology over the past two hundred years. Science has provided the tools that allow "natural" winemakers to make all kinds of "raw" wine today: virus-free vine certification, grafting of European vines onto American rootstocks, anti-fungal materials for disease prevention, inert gas to prevent oxidation, proper production, and treatment of natural cork, gentle extracting wine presses, electricity for operating pumps and processing equipment, refrigeration and temperature control for production and storage, and the biological understanding of sanitation, to name just a few. One could also include wireless computer networks and smartphone technology that allows "natural" winemakers to photograph and post their virtuous efforts on social media. These scientific advances have led to what we enjoy today—the most delicious wines in human history.

As mentioned earlier, I believe making wine is akin to raising a child. The birth is difficult and takes a great deal of effort, while parenting occurs over the long haul. Good parents will allow their children to develop and learn through a child-centered form of parenting—allowing them to grow up while we protect and correct them when needed. Making so-called "raw wine" is analogous to poor parenting. Not doing what is best for your wine could be compared to not paying attention to your child—neglecting education, and not ensuring their health, safety, nutrition, rest, and cognitive development needs. Just like the old days. But is that the best we can do? One French wine writer stated that "natural wine producers think of their wine as children." If that's the case, many of these winemakers would have child protective services knocking on their cellar door.

Some believe the way forward is by looking backward, magically thinking that the way things used to be, was better. It's another version of the romantic notion of the "good old days"—not dissimilar to the philosophy that believes America in the 1950s was the pinnacle of human existence. This narrative also runs through the philosophy of biodynamics, which has been embraced by growers and winemakers around the world. The problem is that biodynamics has also not fared well under scientific examination.

Based on a series of lectures given by the Austrian theosophist Rudolf Steiner in 1924, biodynamics is rooted in anthroposophy, a belief system that seeks to use mainly natural means to optimize physical and mental health and well-being. Biodynamic winegrowing views a vineyard as one holistic organism with the idea to create a self-sustaining system.[3]

As a skeptic of science, it's no surprise that Steiner integrated astrology into his philosophy of biodynamics. Steiner insisted that cosmic influences shaped both plants and humans and believed his approach would provide the "right thing" for the soil. He touted the presence of rhythms and cycles of the earth, sun, moon, stars, and planets and thought these influences and the wider cosmos affected the growth and development of plants and animals. As Steiner was never a farmer, he had no experience growing plants, let alone grapevines. These ideas have been studied for decades and have never held up under scientific scrutiny.[4]

Biodynamic certification requires using specific compost and field mixtures consisting of nine preparations made from yarrow, chamomile, stinging nettle, oak bark, dandelion, valerian, and manure. Steiner numbered these concoctions from 500 through 508. The most bizarre brew is number 500, which consists of cow horns stuffed with manure compost and buried in the ground over the winter. According to the Biodynamic Demeter Alliance, spraying this mixture of manure onto plants will bring them "into a dynamic relationship with soil, water, air, warmth, and cosmos to help them develop in a healthy and balanced way." Again, scientific research conducted on these preparations has found them to be ineffective for disease control.[5] Plus, who wants manure sprayed onto grapes that will eventually be turned into wine?

Even more problematic are Steiner's documented opinions on race. His concern about the environment centered around his idea of a superior race, a philosophy that became part of the German nationalistic movement decades later. Steiner disgracefully claimed that Black people were distinguished by an "instinctual life" as opposed to White people, whom he defined as having an "intellectual life." In the early 1900s, he wrote, "To what extent are uncivilized peoples capable of becoming civilized? How can a Negro or an utterly barbaric savage become civilized? And in what way ought we to deal with them?"[6] This and other similarly offensive opinions were part of Steiner's lectures. The ecofascist wing of the Nazi government eventually embraced biodynamic farming as a way for resettled Germans to maintain their health in newly annexed lands. In a 1933 book entitled *A Way to Practical Settlement,* German landscape architect Max Karl Schwarz claimed that biodynamic farms would help "cure" the land by settling ethnic Germans on "virgin" German territory in "uninhabited" Eastern Europe.[7] These associations, along with the complete inability for biodynamics to be scientifically verified, should be enough to give any biodynamic disciple pause.

Over time and presumably, many failures, biodynamics evolved to rely on several tenets of organic farming, such as composting, biodiversity, and cover cropping. Biodynamic certification now includes the same chemical allowances for insect and disease control contained within the National Organic Program, but only if the application of the nine preparations is adhered to. To receive Demeter certification today, a grower must first achieve organic certification. While not part of Steiner's original doctrine, these organic methods have been proven to enhance soil structure, microbial life, and fertility. However, no scientifically accepted effect of biodynamics and its preparations, separate from organic farming, has ever been established. As one renowned plant pathologist at Cornell University once mused, "it's hard to use the scientific method on religion."

While many adherents to biodynamic winegrowing swear by its effectiveness, an argument could be made that these vineyard improvements are due to the specific use of proven organic practices and the simple fact that growers are paying more attention to their vines under the program. Otherwise, these so-called results are based on nothing more than anecdotal evidence and personal bias. Still, it is a good thing to want to farm safely and ecologically while being mindful of the impact on the surrounding environment. I just feel there are better and more effective ways of doing this.

While practicing biodynamics seems like a harmless endeavor that is trying to rid farms of excess chemical inputs, the acceptance of pseudoscience can lead people to distrust real science, making it harder for the public to become critical thinkers. This is precisely what we've seen during the COVID-19 pandemic. Ironically, the sector of the public that believes in the beneficial aspects of biodynamic wine is often the same group that will preach about "following the science" concerning public health and climate change.

Overall, the main problem with "natural" and biodynamic wine is the underlying false narrative, creating an impression that wines not made in these ways are "unnatural" or "unclean." In the process of promoting themselves as a more virtuous and healthier alternative, natural and biodynamic producers inadvertently damage the overall perception of traditional fine winemaking. As of now, there is no evidence that biodynamic viticulture or natural winemaking are any better for the environment or healthier for people than the safe, traditional methods used in careful, sustainable wine-growing. Rather, these false narratives have proven to be effective marketing tools for wine producers looking to separate themselves from the profusion of labels in the public sphere.

While practicing natural winemaking and biodynamic agriculture could be considered a whimsical endeavor, these philosophies have indeed helped guide modern wine growers away from overly manipulated and often soulless academic production techniques. Today's winemakers have a slew of technology and materials available to them. Many additives and manipulations are employed in the mass production of wine made in tremendous quantities for low prices that satisfy a large segment of the buying public. Winemakers looking to advance quality have rejected this industrial approach, separating themselves and their techniques from mass producers.

Yet the best wines aren't made through benign neglect or wishful thinking. The future of winemaking should not be guided by following pseudoscience or mimicking the ignorance of another age. Learning from history and science should result in understanding what was done before and not repeating the same mistakes. The fact remains that most high-quality wines produced worldwide are made by small, family-owned wineries dedicated to producing sustainably grown, terroir-driven wines with minimal intervention. In this model, science and nature can indeed coexist.

To me, the way forward in winemaking is neoclassical—a science-based approach that embraces the best from the past and combines the lessons learned with new ideas. It allows for the clearest expression of terroir and inhibits anything that would alter that goal. It understands that natural yeast and minimal intervention can bring out a wine's true potential and that the judicious use of sulfur remains the safest and most time-honored way to avoid oxidation and spoilage. It understands that careful, proper filtration remains the best way to remove microbes, ensuring microbiological stabilization and preservation while minimizing the use of chemical additives. It knows that laboratory analysis provides insight into a wine's development, helping us enhance quality and taste. And it embraces viticultural advancements that reduce our dependence on pesticides, helping us to farm more sustainably.

By ignoring or rejecting science, we risk living in the past and repeating old mistakes. Instead, I think we should remember President Barack Obama's words, "If you're walking down the right path and you're willing to keep walking, eventually you'll make progress."[8]

Organic Winegrowing

Farming looks mighty easy when your plow is a pencil and you're a thousand miles from the corn field.

—Dwight D. Eisenhower

Making and growing wine safely is something all producers should strive for. It's essential for our environment and wildlife, as well as our people and communities. Most consumers react positively to the term "organic," which has become an appealing descriptor of food and wine products. Indeed, global production of organic wine grew over 20 percent in 2019 and 2020, significantly higher than non-organic wines, and is predicted to expand at a compound annual growth rate of 10 percent annually in the coming decade.[1] During the five years from 2012 to 2017, organic wine consumption nearly doubled.[2] The appeal of organic wine is clear, but do people really understand what it is? Today, if you ask someone what organic wine is, they'll probably tell you it's grown and made without pesticides or chemicals. But like many things, the truth about organic wine is a lot more complicated.

People have been farming organically since the dawn of human history. They just didn't have a name for it. Only after populations increased and larger, mono-crop plantings became more common did pest control strategies come into play. The earliest documented use of pesticides was about 4,500 years ago in ancient Sumer, where farmers used sulfur dust to protect their grain crops from insect damage. Later, the Greeks and Romans used smoke from burning trash and compost to supplement sulfur to try and ward off insects. They had no idea that sulfur was also protecting their crops from fungal diseases.[3]

Before the Age of Exploration and the Columbian Exchange, most farmers didn't worry much about insects and diseases from other places. Once these pests began to travel around the world, agriculture had to work hard to keep up. Aside from sulfur, farmers used arsenic, wood ash, oil, salt, mercury, lead, and nicotine to protect their crops. With the advent of the Industrial Revolution and, particularly, WWII, chemical research began developing new synthetic materials that farmers could use to help with the growing demand for food worldwide.

The beginning of the organic movement in the West is attributed to English botanist Sir Albert Howard, who studied local farming systems while working in India in the early 1900s. Howard was the first westerner to observe and write about organic agricultural techniques and researched the effects of composting on soil health. He was an early adherent of sustainable agriculture, along with Rudolf Steiner and Eve Balfour. As a trained scientist, however, he was not convinced of the efficacy of biodynamics.

In his seminal book entitled *An Agricultural Testament*, Howard wrote in his preface that "some attention has also been paid to the Bio-Dynamic methods of agriculture in Holland and Great Britain, but I remain unconvinced that the disciples of Rudolf Steiner can offer any real explanation of natural laws or have yet provided any practical examples which demonstrate the value of their theories."[4] Indeed, Howard's opinions regarding biodynamics is reiterated by many of today's agricultural scientists. However, the research behind organic principles is sounder, as methods involving soil fertility, composting, cover cropping, companion planting, crop rotation, and natural pest control are scientifically proven to be beneficial to both farms and the environment.

The modern organic movement in the US began in the 1960s as concerns grew over chemical pollution in the environment. The US National Organic Program (NOP) started in 1990 when the USDA was tasked with developing national standards for organic products. Organic grape growing falls under the same exact requirements as other crops in that synthetic fertilizers and herbicides are not allowed for certification. The program does allow pesticides to be used, albeit from a limited menu of materials. The main materials allowed for disease control are sulfur and copper-based fungicides like copper sulfate and copper hydroxide. These substances are allowed as it's difficult to consistently harvest a crop of grapes without using at least some of these materials at various points during the growing season. The North Fork is like much of the EU in this regard.

The main difference between the NOP and organic programs in the European Union is that the EU allows up to 100 ppm of sulfite into wine while the NOP does not. To use the USDA organic seal on a wine label, all ingredients going into these wines, including grapes and other additives, must be certified organic. Sulfites can't be added, but naturally occurring levels are permitted. Only these wines may display the USDA organic seal.

The USDA has a secondary labeling option called "Made with Organically Grown Grapes," which resembles the EU program in that wines from these grapes can be made with up to 100 ppm of added sulfites. Wines carrying this designation must be made entirely from certified organic grapes and produced and bottled in an organic facility, but they cannot use the USDA's organic seal.[5]

Despite the common perception, organic programs worldwide allow for the use of chemicals like copper and sulfur-based fungicides, a necessary strategy to control downy and powdery mildew. Although the USDA program does not allow for the use of synthetic herbicides to control weeds, organic growers can use synthetic formulations of sulfur and copper sulfate, copper oxychloride, copper hydroxide, and paraffinic mineral oil for disease control. While sulfur and copper are natural elements found in the earth, none of these substances are produced naturally for agricultural use; they are derived from by-products of the oil refining industry. While surprisingly regarded as safe by the NOP, these materials can cause their own set of issues for people and the environment.[6]

Although it's listed as a certified organic fungicide, copper can be toxic to birds and aquatic life. With a chemical half-life of about 2,600 days, copper can become permanently bound in the vineyard topsoil, and its repeated use over many years can create high concentrations, negatively affecting earthworms and other soil microbes. Organic vineyards in the EU have already begun to deal with this issue.[7]

Veteran winemaker and former Cornell research viticulturist Larry Perrine has done extensive work looking into the toxic effects of copper through an investigation of the scientific literature. He has stated that "copper compounds are the single most toxic fungicides used on grapevines and should be avoided. Ultimately, they permanently toxify the soil to such a degree that microbial life is stressed, microbial biomass significantly reduced, and earthworms simply leave the contaminated soil and cease their essential recycling work." From the current research available, Perrine believes that "if organic and biodynamic certification continues to rely on copper to control downy mildew and deny growers the right to use softer materials that don't accumulate in the soil, and are not toxic to microorganisms and earthworms, there is something seriously wrong with these programs."[8]

Sulfur is another organic fungicide that is used mainly to control powdery mildew. As previously mentioned, it's been used for thousands of years but can cause health issues. I've spoken to more than one vintner who has moved away from using sulfur-based fungicides as they can irritate the skin and mucous membranes of the eyes, nose, throat, and lungs. Vineyard workers who tend vines daily are the most affected. Paraffinic mineral oil can be used instead for powdery mildew control, but this material is considered less effective, and when applied late in the season, can sometimes work to delay ripening. There is also some debate about whether mineral oils are potentially carcinogenic to humans while the extensive research on sulfur has shown it to be safe.

Striving to attain organic certification is commendable and important. It isn't easy to achieve, and the rewards are many, but there is vast room for improvement in the NOP program as it's currently constructed. Sulfur and copper-based fungicides must be repeatedly applied in much more significant quantities than synthetic materials to control mildew, especially in wetter climates. A case could be made that using some synthetic materials in lower amounts could be an eco-friendlier approach under these conditions. However, the NOP does not allow for this. Also, proponents of the program have consistently embraced the "no pesticides" or "chemical-free" narrative, failing to correct this misconception for the public. Countless news outlets, journals and websites promoting organic wines describe the program as "growing grapes without chemicals and pesticides," which is patently false.[9] Rather than educating people, this misinformation only makes us less intelligent, opening the doors to further distrust of science. Allowing and encouraging the dissemination of incorrect information only damages the reputation of the program and the spirit of the growers who work hard to achieve organic certification.

The NOP is a national program with no consideration of geographic location and conditions. More than any other crop, growing wine is a highly localized endeavor requiring different management techniques dependent on terroir. The unique strategy of producing terroir-driven wines should behoove the NOP to take local climatic conditions into account. Trying to achieve organic certification shouldn't be a choice between a seal on a label and being truly eco-friendly.

The NOP also needs to be faster to adapt in accepting newer and safer synthetic materials to reduce overall pesticide use. With climate change posing significant modifications to our weather patterns, this issue may become more critical in the years ahead.

The NOP should seriously consider allowing minimal amounts of sulfur in wine, just as it permits its use in vineyards. This would increase the quality of organic wines and create a level playing field in the market against the surge of organic wines produced in the EU. While the US is one of the world leading consumers of organic wine, the vast majority of organically produced wines in the domestic marketplace are imported from the EU. In making this slight change, the NOP would give US producers a more sustainable pathway to organic success.

In an ideal world, the future of organic wine production should look to merge conventional and organic methods with new genetic breeding technology to reduce pesticide use. Unfortunately, the term GMO remains a dirty word, even if the technology has dramatically evolved and improved over the past decades. If we want to ensure progress and meet the growing demand for organically grown and domestically produced wines in a changing climate, we must allow science, not dogma, to lead the NOP into the future.

The bottom line is that all pesticides used in growing grapes, whether organic, biodynamic, natural, or synthetic, have some risk associated with their use. The NOP, imperfect as it is, is the only organic program available for growers looking to be certified in the marketplace. Organic grape growing on the North Fork can be done, but as our weather changes quickly and the seasons vary from year to year, it's difficult to rely on the program for consistent, sustainable success, especially with European wine grapes. The best way forward should be to try to limit toxic materials as much as possible while allowing growers some flexibility to safely navigate the vagaries of weather. That's where the philosophy of sustainability shows its importance.

Sustainability

What's the use of a fine house if you haven't got a tolerable planet to put it on?

—Henry David Thoreau

Most people understand environmentalism and how it applies to the world around them. In the last decade, a recurring meme in the public consciousness revolves around "sustainability." The question on many people's minds is, "What exactly does this mean?" Here's a quick primer that I think will help.

The concept of sustainable agriculture grew out of the early organic movement and became fine-tuned during the late 1980s. The first use of the term was coined by the Brundtland Commission of the United Nations on March 20, 1987. It stated: "Sustainability is development that meets the needs of the present without compromising the ability of future generations to meet their own needs."[1] In 1989, the American Agronomy Society adopted the following definition for sustainable agriculture: "A sustainable agriculture is one that, over the long term, enhances environmental quality and the resource base on which agriculture depends; provides for basic human food and fiber needs; is economically viable, and enhances the quality of life for farmers and society as a whole."[2] Agriculturally, sustainability means growing a crop mindfully, using methods that do not completely use up or destroy natural resources or negatively affect workers and the surrounding communities, thus ensuring the viability of the farm for a long time. By farming with a minimal impact on the surrounding society and environment, we meet the needs of the present without compromising the ability of future generations to meet their own needs. It's a way of life and a pathway to long-term success for us and our environment. In other words, we try to stop fouling our own nest. Sustainability revolves around three main points: utilizing ecologically sensitive practices, socially supportive strategies, and economically viable techniques. Some term this philosophy the three Es— environment, equity, and economics. Where these three concepts interact together is the "sweet spot" of sustainability.

The methods employed for sustainable winegrowing depend on the grape varieties and where they are grown. While sustainable practices are highly localized, organic certification is national, and Demeter is international. As mentioned, these two well-known certification programs may not always succeed at the local level, and they each have their own sets of issues. Sustainable agriculture provides another pathway for growers seeking to become more eco-friendly but who are unable to farm successfully under the arbitrary limitations of the NOP or unwilling to spend precious time and resources on the unproven, incoherent aspects of biodynamics. Instead, sustainability utilizes the scientifically verified practices of organic agriculture while transcending the false choices offered by biodynamics, understanding that farms and people don't operate in a vacuum. Sustainability is not an ideology. Instead, it's practical and research-based—not a government program or a philosophy developed in a far-off time and place. It's founded on a solid scientific approach to deal with the complexities and challenges of agriculture while encouraging us to care about our environment, neighbors, and crops using a system of localized "best practices."

Long Island vineyards have worked hard to develop unique and safe practices for producing quality wine grapes. As part of a fragile ecosystem surrounded by water, local producers have long believed we must be careful stewards of our land. Vineyard owners recognize this not just in terms of the present day, but as an essential factor in the region's long-term viability.

For these reasons and to codify what many growers had already been practicing, a handful of vintners met in the spring of 2011 to begin developing the philosophy, recommendations, and practices that became the Long Island Sustainable Winegrowing program (LISW). With assistance from Bedell owner Michael Lynne and CEO Trent Preszler, we were able to provide the resources to develop the infrastructure of the program and get it off the ground. Today, LISW remains one of the only third-party sustainable certification programs on the East Coast for wine grapes and has grown to include twenty-two member vineyards and over 1100 acres of vines. As of this writing, LISW remains one of only fourteen such programs worldwide for wine grapes.

Every year, members must follow nearly two hundred specific practices that require using safer and less toxic materials, limiting nitrogen fertilization, and incorporating numerous ecological-management options including reducing or eliminating herbicides. Member wineries must plant cover crops, compost their grape pomace, and practice integrated pest management (IPM) to scout for insects and diseases. This allows us to manage vineyards with precise surgical strikes rather than using a blanket chemical treatment with unacceptable side effects.

These practices typically include organic materials and concepts. Yet, as is often the case with wine grapes on the humid East Coast, growers must be flexible enough to manage outside organic dogma to be sustainable. As mentioned, organic viticulture depends on using copper-based fungicides to control downy mildew—one of the most serious diseases of wine grapes. Many in the scientific community believe that some synthetic materials are safer and break down in the soil more quickly than this heavy metal (and organic) pesticide. Unfortunately, "synthetic" compounds cannot be considered "organic." But following sustainable agricultural practices is not about ideology nor just about insect and disease control. It's about managing small farms in a way that can sustain success over a long period in the community they're situated in.

One primary goal of LISW is to conserve Long Island's delicate maritime and estuary ecosystems by protecting our ground and surface waters from leaching and runoff. The program limits nitrogen use and discourages fall fertilizer applications to reduce leaching. Most of what we remove from the vineyard—the fruit—is returned after pressing as compost. The remaining vine parts—leaves, shoots, and canes—are returned to the soil, leaving a small net nitrogen removal. It's hard work to stay ahead of the game, and participating vineyards must be verified yearly through a vineyard inspection, along with a review of grower records.

The concept of regenerative agriculture is also part of the sustainability narrative. Regenerative agriculture describes farming practices that, among other benefits, help to reverse climate change by rebuilding soil organic matter and improving soil biodiversity. The three most important practices of regenerative agriculture are reduced or no-till farming practices, cover cropping, and composting—all of which are required practices in the LISW program.

Growing grapevines is very regenerative as they are a perennial crop requiring little tillage. Soil fertility increases in regenerative systems through cover cropping and compost applications, which restore the plant/soil microbiome to promote the liberation, transfer, and cycling of essential soil nutrients. The LISW program requires permanent cover crops between vine rows, which improve soil health and increase organic matter. It also requires that all pomace, stems, and skins left over from grape processing be composted and returned to the same vineyard where they came from. The inoculation of soils with compost not only helps restore microbes but also improves soil structure.

Building ecosystem diversity is also part of LISW, as member vineyards must establish ecological compensation zones where plants (other than grape-vines) can live and thrive. These zones can include multispecies cover crops and vineyard borders planted for pollinators and other beneficial insects. Under-trellis plantings are also encouraged to reduce tillage and herbicide use. Mechanically breaking up soil negatively affects soil aggregation and can be one of the most degrading agricultural practices, significantly increasing erosion and carbon loss. Minimizing tillage, along with these other regenerative practices, reduces erosion, enhances soil structure, improves water infiltration and retention, and increases carbon sequestration.

Is sustainable agriculture a perfect system? No, and none of the other eco-certification programs are either. But the sustainable model remains the best overall strategy for long-term success. Examples of unsustainable winemaking still dominate the marketplace. Mass-produced, mechanized, large-scale, industrial-style wines, manipulated with additives and made with minimal hand labor, are transported thousands of miles to consumers, leaving a large carbon footprint. This is ultimately not a sustainable system.

LISW was formed to try to make our communities a better place to live and work while producing some of the best wines in the US. Through the LISW program, East End vineyards and wineries have endeavored to create their own definition of sustainable viticulture. The viability of local vineyards depends on our ability to steward the land in a way that allows it to stay healthy and productive into the future. We need to remember the water flows underneath us every day; what we do on our land today will be in someone's glass of water fifty years from now. We need to pay attention.

LISW will help ensure that wine growing remains viable in our region for generations to come. We're on the way to making the sustainable practices required by LISW the gold standard on the East Coast. The program believes vineyards should work in harmony with our natural world, leaving the land in better condition than when we found it and building a community between vineyards, workers, and the land. Not only will it help protect our sole source of drinking water and improve the environmental conditions of our surrounding waterways, but it will also help us to make better and more distinctive wines.

Sustainability starts at home with all of us. Recycling, conserving energy, and supporting your local farms are all sustainable actions we can take on our own. We need to be open to buying products emphasizing the social, environmental, and economic benefits of paying a fair price for locally grown produce. Certified member wineries of LISW use a "certified sustainable" seal on their labels. This logo can only be used by vineyards that have completed and passed a rigorous inspection. Wine labels carrying the LISW seal must also contain at least 95 percent sustainably certified fruit. It's important to look for it; when you see it on the bottle, you'll know that the grapes were grown under the LISW program.

Consumers are increasingly learning about the impact of their food and wine choices on their total quality of life. So next time you want to pull a cork, remember all that goes into your local wine—and think about sustainability.

For more information on Long Island Sustainable Winegrowing, visit www.lisustainablewine.org.

LISW Certification Seal. Courtesy of lisustainablewine.org.

Flaws

Experience is simply the name we give our mistakes.

—Oscar Wilde

Sometimes in wine, as in art, beauty is in the eye of the beholder. As the old cliché goes, great wine is made in the vineyard. In most cases, a flawed wine is made in the cellar. As a winemaker, one of my main jobs is trying to prevent flaws from occurring. Most flaws in wine are allowed to happen through a lack of attention. Grapes and wine, under the proper care and guidance, really should be flaw free.

While vineyards are the place where wine is essentially made, they are also the place where wine flaws can begin. Poor vineyard management or bad harvesting decisions by a vintner can affect otherwise healthy fruit. Without some form of initial intervention in the cellar, these qualities can translate from the vineyard into off-flavors in the bottle.

Many wine flaws can also arise from negligence in the cellar. Oxidation and acetic acid (vinegar) can occur from rough handling and insufficient sulfur, while reductive odors can be produced from a lack of wine movement or racking. Other issues can show up during the bottling process; wines poorly handled during bottling sometimes develop problems months later like re-fermentation or unwanted Malo-lactic growth, creating off-putting textures, flavors and aromas for the end consumer. Keeping wines clean and stable is the winemaker's job from the day the fruit arrives in the cellar until the day of bottling. It's not as easy as it appears, and a good winemaker must stay vigilant. It's not about intense manipulation or over-involvement. It's about being mindful and providing the conditions for a wine to be all it can be—like parenting.

One of the ways that wine flaws can be addressed, whether originating from the vineyard or cellar, is the careful use of fining. Fining is a centuries-old process to clarify wine and is also used to remove unwanted flavors and aromas. Most fining agents work by attracting the positively or negatively charged matter suspended within a wine, causing these particles to bind to the fining agent and precipitate to the bottom of the vessel. After a short period of time, the wine is then racked or filtered to remove the remaining fining agents and the characteristics bound to them. Some common fining agents used throughout the world include gelatin, isinglass, egg whites, casein, bentonite, and carbon. Gelatin and egg whites have a positive charge and are used to remove negatively charged tannins, "softening" wine and reducing its astringency or bitterness. Bentonite, a clay from volcanic ash with a negative charge, is commonly utilized to remove excess, positively charged proteins that can later produce a cloudy haze in the bottle.

Other fining agents like carbon and isinglass are nonspecific and used mainly as overall fixes for more serious issues. As a certified vegan winemaker, I avoid the use of animal-based fining agents like eggs and isinglass. Modern technology has developed hundreds of synthetic fining agents over the years for winemakers to use. I believe it's best to not use any fining at all to preserve terroir, but I'll often use bentonite clay to remove excess protein. Most reasons for fining can be avoided through careful vineyard management and judicious winemaking practices in the cellar.

Fining differs from filtration. Most fining agents remove flaws through chemical means while filtration is used to physically improve wine clarity and later, to separate unwanted microbes from the wine (yeasts and bacteria) prior to bottling. I prefer to sterile filter all my wines prior to bottling as I believe the removal of yeast and Malo-lactic bacteria produces wines that will taste delicious and flaw-free for decades.

Both techniques are important tools for a winemaker to improve wine quality and stability. They should be used carefully and carried out only when necessary as it is possible to remove positive flavor characteristics. Like many cellar procedures, fining and filtration are only as good as the winemaker using these techniques.

Of course, the definition of a flaw can sometimes be subjective. Many great and expensive wines show symptoms of *Brettanomyces* (known in winemaking circles as "Brett"). It can be a subtle nuance, or it can overpower the authentic flavors of the wine. Some folks will say this is terroir, but I disagree. No matter where it's found in the world, *Brettanomyces* always manifests itself in the same ways, creating a barnyard-like, *Band-Aid aroma,* masking the transparency of the vineyard and the area from which the wine came. Most of the time, its presence is due to poor cellar maintenance. Ironically, if Brett is found in wine from one of the world's most famous and expensive Châteaux, it's considered an exquisite quality; if found in a North Fork wine, it's often considered a flaw, although small amounts can add some interesting complexity.

Aside from carefully timing the harvest and gently handling fruit, the most critical tool in the winemaker's toolbox to control flaws in the cellar is the careful use of sulfites. It's one of the oldest techniques we have in making fine wine, yet its reputation has suffered from inaccurate and misleading information, both inside the wine world and in the mass media. Sulfites used judiciously in the correct amounts, and at the right time, can prevent oxidation and browning, unwanted vinegar bacteria, and *Brettanomyces*, all while preserving the varietal character of the fruit. Its use is among the safest and most time-honored techniques in winemaking.

Naturally, the overuse of sulfite and its overt presence in a finished wine can also become a flaw. That's why winemaking is both art and science. You need to understand both if you want to be a great winemaker.

Sulfites

Countless pleasures are wasted through ignorance and a want of skill and attention.

—Emile Peynaud

You know all those urban legend stories—like the one about alligators in the NYC sewers and the blind date that ends with you waking up in a bathtub of ice. The wine industry has its share of urban legend stories, with one that I still hear every year. It goes something like this.

A couple returns home from a recent trip to Europe. The couple talks about how they drank wine like crazy and never got tired or had a hangover. They reminisce about how they met local winemakers who told them, "American wines all have sulfites, and ours don't." The couple agrees that when they are home in the US, they cannot drink wine the same way and enjoy it as much. They insist that the sulfites used in American wine gives them a headache, and they inevitably want to know why we use it. They usually stick to white wine "because of the sulfites in reds." They never buy another bottle of American wine.

My answer to them is always the same: "Perhaps it has something to do with you being relaxed and on vacation!" I say that the reason they feel so good in Europe is simple; they are away from home and the kids and the daily grind of work or maybe they're sleeping a little later and having lots of "intimate time." At this point, most of them give me dirty looks and shake their heads in disbelief. Some of the wives will snort and elbow their husbands in the ribs. I try to explain that there's nothing in the medical literature proving sulfites have anything to do with headaches and that red wines contain lower sulfites than whites. By now, of course, I've lost them.

Over the past decade, sulfites have become some of the most vilified, controversial, and misunderstood compounds in the wine world. They've been blamed for everything from headaches to congestion and subsequently avoided by consumers who believe this false narrative. Though most industry professionals understand that sulfites don't cause headaches, the debate around them and their positive and negative effects on wine continue.

Whatever people want to believe, one thing is for sure; the problem is not sulfites. All European wine producers use sulfites in wine production—in the vineyard and the cellar, and all wines contain sulfites, whether added or not. Wines without any added sulfites at all can still contain anywhere from 5–40 ppm. Typical levels in finished, bottled wines range from 5–30 ppm. The yeasts that convert sugar into alcohol also produce sulfites as a by-product. The human body produces about 1 gram per day. Years ago, the subject of sulfites in European wines never even came up; the EU has only been required to list sulfites on the label since 2005. Many imported wines can contain higher levels of sulfites than domestic products. European wineries are allowed to use far more additives than we are in the US. They invented additives for wines.

Chances are, you will ingest more sulfites in your average restaurant dinner than a glass of your favorite wine. Dried fruit, French fries, packaged meats, prepared soup, soda, frozen juice, shellfish, soy flour, maple syrup, prepared guacamole, sushi, olives, pizza, cheese, crackers, and fish (the list goes on) can contain more sulfites (in milligrams per liter) than most wines. The average bag of dried fruit and nuts contains more than ten times the amount of sulfites found in a bottle of wine. Why doesn't anyone ever complain about trail mix giving them a headache?

Sulfites in domestic wine was not an issue until the mid-1980s when a few anti-alcohol legislators lobbied the FDA to develop a warning label for wine containing more than 10 ppm. Some European producers saw this as an opportunity to set themselves apart from their up-and-coming American competitors. Don't be fooled. Most imported wine contains more sulfites than domestic bottles in order to stabilize it for ocean shipping.

Just so, people who are allergic to sulfites must be very careful. The most dangerous sulfite reaction involves anaphylactic shock that constricts the breathing passages and severely lowers blood pressure. This type of reaction only occurs in about 0.4 percent of the total population or about 150,000 people. If you have asthma, there's a 5–10 percent chance you'll have a sensitivity to sulfites. In comparison, about 4 percent of the population (about 11 million people) suffers from severe food allergies. As an example, peanuts are far more dangerous than sulfites can ever be. Since 1990, the FDA has reported nineteen sulfite-related deaths—none from wine. Most of them were from prescription drugs containing high levels of sulfites (200 ppm and higher). Peanut allergies alone result in at least one hundred deaths per year.

As I tell my customers, you shouldn't worry about sulfites in wine unless you are one of the few people that are truly allergic. If you want to worry, there is something in wine you should be concerned about. Alcohol is a well-studied neurotoxin and psychoactive substance with dependence-producing properties to the human body. It's also been proven to be a known carcinogen at high consumption levels. It typically makes up 10–16 percent by volume of an average bottle of wine. What do you think has a greater chance of causing you harm—20 parts per million of sulfites or 14 percent alcohol?

According to the National Council on Alcohol and Drug Dependency, more than 140,000 people in the US die annually from alcohol-related causes, including falls, car accidents, and liver cirrhosis. Add to this the tens of millions affected by alcohol-related illness and addiction. This is sobering information, but part of enjoying and appreciating wine must include respecting it and practicing moderation.

For over four hundred years, European wine producers studied the effects of sulfites in wine. They learned they could not make good wine without its use. We learned everything we know about it from them and have continued to improve our knowledge. European wines contain far less alcohol than most New World counterparts; everyday table wines found in the Old World contain even less alcohol than their exported versions. It's why tourists can enjoy imbibing wines in the Old World without as many annoying after-effects. These same tourists would find that North Fork wines have the same qualities and provide a similar experience.

As a society, we tend to react negatively to the awful-sounding names that science has given to ordinary, natural things, many of which have been around far longer than human beings. The goal of science is to examine, identify, and find the truth.

We need to do a better job of stepping back and understanding the bigger picture. And the next time you're in Europe on vacation, remember to enjoy yourself, drink lots of good wine, and set those little old wine-makers straight.

Zen and the Art of Winemaking

Chance favors the prepared mind.

—Louis Pasteur

Mindfulness generally, and meditation, specifically, lets us regain control of our minds and our emotional state of well-being. It permits us to be fully present, aware of where we are and what we're doing, and not overly reactive or overwhelmed by what's happening around us.

Being mindful means, the questions we ask are more important than the answers we come up with. The goal is simple: we aim to pay attention to the present moment without judgment.

I feel that mindfulness has a real place in producing quality, terroir-driven wines. Like every other part of life in the digital age, the world of wine production is full of information, guidelines, and opinions, which can all lead to distractions and misdirection. Mindfulness is essential when tasting wine, where close attention is paid to fine details of aroma and flavor. For this reason, I thought I'd include my thoughts about mindfulness in winemaking and how I experience it.

- The best things you can leave behind in any vineyard are your footprints.

- We make wine on the land where we live. Be aware that everything we do in the vineyard and cellar will be reflected in our environment, both above and below the ground.

- Give the wine permission to be what it wants to be.

- Stressing and worrying about the weather does nothing. It is out of our control. Accept the weather, embrace it, and guide your winemaking accordingly.

- Fermentation is a wonder of nature. Rejoice in what it can do and guide it the best way you can.

- One should never try to make a wine with a distinct style in mind before really comprehending the vintage and tasting the fruit to see what it's capable of.

- What does the wine want to do today? What do you as a winemaker want to accomplish today?

- As soon as one tries to change what a wine wants to become, one begins to fight back against nature. Be confident that wine reflective of terroir will be unique and delicious.

- Sometimes, the best thing a winemaker can do to wine is nothing. It's best not to be overly reactive.

- Observe the seasons and have confidence in your terroir; try to feel what the vines have been going through to get their fruit to ripeness. Listen to the vintage.

- Mindfulness in winemaking doesn't necessarily mean making organic or natural wines. Try to let ideologies go; instead, work with the vines, be aware of the season, and accept the weather you receive and the steps you need to take to guide the wine into the bottle.

- Avoid self-criticism and judgment while identifying and managing difficult conditions. Try not to be yanked around by your emotions.

- Wine is made by people. We need to grow and nurture people just as we do the vines that produce our fruit.

- Winemaking has become inundated with apps, laboratories, additives, and social media like the rest of our lives. This can be anxiety-producing and create a sense of disconnection. Try to understand that none of these are necessary to produce and enjoy wine.

- When tasting wine, stay present with the sensory experience happening inside you and gently focus your attention on what your palate is feeling. Trust yourself.

- Drinking and experiencing the pleasures of wine doesn't have to be an intimidating, academic experience. Throughout history, wine was meant to be enjoyed and help us celebrate life. Drink it. Enjoy it.

Vineyard Meditation. Courtesy of Steve Carlson, Bedell Cellars.

PART II

SPONTANEOUS FERMENTATION

First Harvest, c. 1983. Author's personal collection.

Climate Change

If you do not change direction, you may end up where you are heading.

—Lao Tzu

It's rare when a day goes by without another story about the possible consequences of climate change. Many world leaders have called it the greatest challenge of our time. Although the exact outcomes and causes are disputed, few now doubt the phenomenon's existence. The Pentagon predicts climate change will be the most significant destabilizing factor affecting US security in the future. Many people would like to know what climate change will mean for us and our way of life. As a winemaker, I understand that climate is critical to a wine region's sustainability. So, what kind of changes can we expect?

When I was a graduate student in science education at Stony Brook in the late 1980s, I did my master's thesis on climate change and its potential effects on Long Island. In those days, it was termed "global warming," and most people didn't take it too seriously. But even back then, the library shelves were full of research journals on the topic, and it soon became clear to me that climate change was something we all needed to pay attention to.

Back then, the weather models were not as accurate as they are today, but scientists agreed that the main issue for the East Coast was sea-level rise and the potential for flooding along Long Island's coastline. I made a transparency map showing the decrease in Long Island's coastal areas as the sea level increased. While the time frame may have shifted, these models still hold today. Recent studies in 2021 by the NASA Sea Level Change Team at the University of Hawaii predicted more frequent high tide flooding on US coastlines by the mid-2030s, exacerbated by climate change-induced sea-level rise and tidal patterns influenced by the moon's 18.6-year cycle.

117

The NASA study predicts a minor increase in flooding for the Long Island region in the late 2020s, followed by relative stability until the early 2040s when flooding will increase.[1]

Researchers from Stony Brook University agreed that the NASA study could result in more coastal flooding as storms increase in severity. Kevin Reed, a Stony Brook University professor who studies climate modeling and extreme weather events, stated, "Storms are going to be more intense by 2030, meaning they could have larger storm surges associated with them."[2] While climate models are complicated and are constantly updated, I do find it interesting that the ominous predictions for coastal flooding have not changed much in the last thirty years. It's an issue our local government needs to take seriously.

One of the first scientists to discuss the problems of climate change on wine growing was Dr. Anthony Del Genio, a planetary physicist. He also worked at the NASA Goddard Institute for Space Studies (GISS) at Columbia University. He's been studying climate models since the 1970s and is also a wine buff. In 1984, Dr. Del Genio published "Earth's Changing Climate: How Will It Affect Viticulture?" He wrote: "In general, regions now considered ideal for wine grapes may become too warm, while those marginally warm enough for fragile *vinifera* will begin to flourish. Maritime climates should be influenced the least since the oceans will remain cool due to the partial melting of the polar ice caps."[3] Most importantly, he stated that "all of us will begin to feel the effects of climate change by the end of the 1990s."[4] He was not only prescient, but he also had some excellent data. Needless to say, he was way ahead of his time.

For the past few decades, Dr. Del Genio and his colleagues have studied the relationship between precipitation (P) and evapotranspiration (ET). ET is the sum of water loss from soil evaporation and plant transpiration. He described the tension between water supply and demand as precipitation (P) versus potential evapotranspiration or PET. The emerging pattern in the Northeast is that rainfall will increase with warming, but PET will increase more quickly, resulting in a net water loss. This would mean that, over time, our aquifer could be significantly stressed. How quickly depends on how much the climate warms. Dr. Del Genio stated, "If we do nothing to decrease CO_2 and other greenhouse gas concentrations, we might expect 3°C–6°C warming by the end of the 21st century in the Northeast. If we can cut emissions by a significant but modest amount, we may get half that much warming." The message is clear; we all must change our habits to conserve the water supply for future generations.[5]

Although Long Island will experience slightly more precipitation over the long haul, more water will evaporate out of the soil than rain will fall on it. With less frequent but more intense rainfall as the climate warms, more water will go into runoff, and less will infiltrate into the soil, where it can reach underground streams and aquifers. This makes conservation, as well as changes in public policy, even more important.

Growing Degree Days is one of the main tools used to track climate change in agriculture. As mentioned earlier, this system cumulatively calculates the average daily temperature of the growing season using a base of 50°F. Winegrowers worldwide use this and similar systems to define the potential of their climate and the types and varieties of grapes they can successfully grow.

I decided to look at some more recent data from the past twenty years and see if anything new turned up. For one, Northern California is becoming much warmer. Napa, which used to be classified as a Region II back in the 1970s, is now experiencing seasons well into Region IV. Other areas within Napa Valley and Sonoma have reached well over four thousand GDD (Region V) in the past few years—approaching conditions that might make it difficult to produce quality wines in the future. California wine producers are taking these predictions seriously, as seen by the numerous land purchases by larger wine corporations and viticultural speculators in the cooler areas of the Northwest. The numbers have risen so much that many California producers have adjusted and changed the parameters of traditional GDD and are using several different systems to downplay the temperature increases.

Back East, things get a little more complicated. Data obtained for the Finger Lakes region since the 1970s show no statistically significant increase in GDD. It is surprisingly consistent, with average GDD hovering around 2700–2800 for most areas around the lakes.

Eastern Long Island presents an entirely different story. When I first wrote and applied for the two local AVAs—*The Hampton's, Long Island* in 1984, and *The North Fork of Long Island* in 1985, I used data that went back to the 1940s. According to that information, the North Fork averaged 2932 GDD while the Hamptons averaged 2531 GDD. Over the past two decades, the numbers have looked quite different. Since 2000, the average GDD for the North Fork is 3407 days, while the average for the Hamptons is 2960—an increase of 475 and 429 days, respectively, for these two regions. Although it is not a long enough time for a definitive conclusion, one can see the trend; on Long Island and in California, average increases of over 400 GDD have been recorded since the 1970s. This is statistically very significant.[6]

I recently looked at rainfall data on the North Fork; the information I found went back to 1938. Average monthly precipitation from March through October remained steady at 28.5 inches until the early nineties. From 1990, precipitation during this same period averaged 31.5 inches. Interestingly, the number of rain events declined while the total amount of precipitation increased. This means we're experiencing fewer storms with more rain per event. Concurrently, average temperatures rose significantly, similar in some ways to the European studies. What makes the North Fork different from Europe is that our precipitation occurs earlier in the growing season; the months of August and September seem to be getting drier.[7]

I contacted Dr. Del Genio again to get his thoughts on what has changed in his understanding since 1990 and what he sees in the future. Interestingly, he mentioned that he and his staff had begun drawing new conclusions from the models used to predict climate change. He stated that since the onset of clean air legislation that began in the 1960s, the levels of particulates (e.g., dust, soot) in the air have decreased, which ironically has led to higher recorded temperatures in the densely populated areas of the Northeast, Midwest, and California. These particles in the air, known as aerosols, scatter sunlight and make the light less concentrated—in essence, masking the warming effects of elevated CO_2 levels. The areas with the greatest response to clean air legislation show the greatest temperature rise today—as much as 2°F–3°F on average.[8]

This effect is seen on Eastern Long Island since we are close to New York City. But Dr. Del Genio reiterated his earlier opinion that "Eastern Long Island should be influenced the least since the oceans will remain cooler than land because ocean currents store some of the heat at the bottom of the ocean."

Our reduced aerosols and slight increases in water temperature are more than likely responsible for the increase in GDD. It also explains why a region like the Finger Lakes, further removed from air quality issues and with more of a continental climate, is presently showing less of a warming effect. This is one good example of how complex our climate system is and why it is so hard to accurately determine future outcomes.

As for the future, Dr. Del Genio stated that new climate models predict we will experience increases of anywhere from 400–500 GDD over historical averages by the end of the decade. I mentioned to him that I had run the numbers and saw this beginning to happen. He was not surprised;

again, Dr. Del Genio was on the money. When I asked him what stocks he would recommend, he said he was only a physicist.

Other studies done in 2011 by climate scientists at Stanford University looked at how the change would affect the California wine industry. They reported that by 2040, the average temperature in Napa could increase as much as 2°F, reducing the amount of land suitable for cultivating premium wine grapes in Northern California by 50 percent. They stated that "as much as 81 percent of premium wine grape acreage in the country could become unsuitable for some varieties by the end of the century."[9]

The first worldwide scientific analysis regarding climate change and wine production was completed in 2013 by Conservation International. Their key finding: climate change will dramatically impact many of the most famous wine-producing regions in the world and prompt the opening of new areas to wine production. "The fact is that climate change will lead to a huge shakeup in the geographic distribution of wine production," said Lee Hannah, a senior scientist at Conservation International. The study forecasts sharp declines in wine production by 2050 in Bordeaux and the Rhone regions in France, Tuscany in Italy, Napa Valley in California, and Chile, because it will be too hot. Dr. Hannah predicted quality wine production would increase in places like northern Europe, Britain, and the East Coast of the US and that wine regions would develop in areas once considered unsuitable.[10]

While most of the past research has centered on warming temperatures, scientists today are beginning to look more closely at precipitation. Research led by Dr. Ben Cook of NASA's Goddard Institute for Space Studies at Columbia University confirmed that climate change is diminishing an important link between droughts and the timing of wine grape harvests in France and Switzerland. In a 2016 paper published in *Nature*, titled "Climate Change Decouples Drought from Early Wine Grape Harvests in France," Dr. Cook stated, "Wine grapes are one of the world's most valuable horticultural crops, and there is increasing evidence that climate change has caused earlier harvest days in these regions in recent decades."[11] By looking at wine grape harvest dates from 1600 to 2007, Dr. Cook and his colleagues discovered harvest times began shifting dramatically earlier during the latter half of the twentieth century. Early harvests from 1600 to 1980 occurred in warm and dry conditions; from 1981 to 2007, earlier harvests occurred even in years without drought. Their research showed that in the past, you'd only get a warm season when it was dry, but now, warm conditions exist whether it is dry or wet. Whether high-quality wines can be produced during these warmer, wetter seasons remains to be seen.[12]

Research from the University of California Davis and Bordeaux Sciences Agro showed that both regions had warmed substantially over the past sixty years. This warming has contributed to increases in the average wine quality. The studies showed that "Napa and Bordeaux are warming at a remarkable rate" and warn that "ripening relationships revealed that we are reaching a plateau and raise concerns that we may be approaching a tipping point in traditional wine-growing regions."[13]

Opinions among local wine veterans raised many concerns for the industry's future on the East End. Alice Wise, the research viticulturist with Cornell Cooperative Extension of Suffolk County, believes climate change "is and should be a topic of great interest to all those in agriculture." She states that "from a purely selfish viewpoint, slightly warmer summer temperatures would benefit late-ripening varieties. However, we all share the same concerns about how warmer temperatures affect all aspects of the environment, particularly the marine waters that moderate our weather so profoundly. My biggest concern is the potential for more frequent and intense tropical weather in the post-veraison period." Ms. Wise is also worried about elevated ozone levels, which can create leaf injury and might be an even greater issue in the future with increasing summer temperatures.[14]

Larry Perrine, who has grown grapes and made wine on both forks of the East End, worries about the potential for rising sea levels. "If sea-level rise occurs according to some of the models, it presents a big problem in coastal areas," he said. "What about the rainfall? When rain increases, then disease pressure does also. Disease-resistant grape varieties become more important." Conversely, Mr. Perrine states, "If the weather becomes warmer and drier, it could make the area more conducive to red wines." Given a choice, the latter scenario is one I'm sure most local wine producers would prefer.[15]

My dear friend, the late Ben Sisson, a vineyard manager for over thirty years on the North Fork, was also concerned about the uncertain effects of climate change on precipitation. "I think all vineyards will have to have irrigation available, as the dry stretches seem to get longer each year," Sisson said. "I don't think it is necessary to water regularly, but the ability to get water to the vines if they need it will be essential." Mr. Sisson said that the real difficulty in dealing with managing vineyards lies in the age-old uncertainty of weather prediction. "Vineyard managers have to deal with a crop in constant flux, and many of their decisions are based on experience gained from past seasons," he said. "If those parameters change, it will make growing grapes much harder."[16]

Temperature is a critical factor for grape growing. An uptick in heat causes disparities in the color and aroma of the grapes and the accumulation of sugar and acid levels in the growing process. A warmer climate could lead to the dilution and disappearance of terroir in certain regions and some traditional wine styles we are familiar with—particularly those in the cooler wine-producing regions of Europe. Dr. Robert Pincus, a climatologist at the National Oceanic and Atmospheric Administration (NOAA) in Colorado who has written extensively about the wine industry, states in the journal *Gastronomica* that "in an increasingly warm world, the particular associations between wine and place will be difficult or impossible to maintain."[17] His studies led him to conclude that "even where the impact of climate change is less dramatic, decades, even centuries of viticultural experience will be rendered irrelevant."[18] Aside from higher temperatures, climate models predict changes in the levels and rates of precipitation and, most importantly for Eastern Long Island, stronger and more frequent hurricanes, which could make life a little more precarious for everyone—winegrowers included.

Another interesting phenomenon we've witnessed over the last decade is the increasing importance of ultraviolet (UV) light penetration. As an example, we saw almost identical levels of warmth (calculated as Growing Degree Days) in 2012 as in 2011—but the two years couldn't be more different. The 2011 vintage is considered by most local vintners to be one of our most challenging in terms of weather. While the reds from 2011 were more delicate and less extracted, the 2012 reds generated higher sugar levels and showed much more intensity and power. The difference was that 2012 had a far greater number of clear sunny days than 2011, as well as much less rain. During the growing season (March 1–October 31) of 2011, we accumulated about 35 inches of rain, while in 2010 we had 28 inches. The 2012 vintage had 25 inches of rain; it's the lowest of the three years yet is not considered a great vintage. The lesson here is that great vintages are not just about heat, but the confluence of heat, sunlight, and dry weather that can lead to truly extraordinary wines.[19]

We've already experienced the effects of warmer North Fork winters on our vineyards. A big concern surrounds the persistence of insects and disease organisms. Historically, populations of these pests were usually reduced by cold winter temperatures. Likewise, the local tick population could increase with less severe winter temperatures, forcing our vineyard workers to become more vigilant about their safety.

A more positive effect of milder winter temperatures is the effect on yield. Over the last ten years, we've had a few extremely mild winters, followed by substantial crops—some approaching five tons or more to the acre before they were thinned. It's no coincidence. In most North Fork winters, the temperature drops low enough to cause some minor bud damage. The overwintering buds contain embryonic grape clusters; when low winter temperatures arrive, these buds are slightly damaged, reducing overall yield. This is one of the reasons why most New World vineyards (especially on the West Coast) can consistently produce high yields of quality fruit at six to eight tons per acre. This may be a positive for some varieties; however, North Fork vineyard managers must remain diligent in their crop control, especially when looking to ripen late-season reds for high-quality wines. The qualitative target for growers tends to remain at around three tons per acre. The winters may be warmer, but we're still a cool, maritime climate.

In the future, a warming climate may mean many things. On the positive side, it may extend the length of our growing season on the North Fork, increasing our ability to ripen late-season red grapes. It may even open entirely new areas for viticulture on sites where vines have not been previously successful. In conjunction with a decrease in late-season rains, this would provide us with even more great vintages. Grapevines need less water than most other crops and are well suited to withstand long periods of drought. As our vines continue to age, root systems will grow even further into the subsoil to find moisture younger vines can't reach. For these newer vines and vineyards planted on very arid soil, drip irrigation becomes even more important during extended dry periods.

For winemaking, climate change means that terroir will also change, resulting in subtle effects on wine style. For the time being, most of these modifications will be positive; however, we must remain mindful that this benefit may be temporary. Luckily, our maritime conditions buffer against climate change's more extreme effects. As a cool climate region, warmer temperatures are welcomed and water is not an issue, at least for the foreseeable future. Even with the possibilities of more severe weather events and hurricanes, the North Fork will continue to have some of the most moderate weather in the country. We are in much better shape to brave the future than the wine-producing areas of the West Coast.

The term we need to use is not global warming, but climate change. As the name states, climate change is just that. It's not only about warming; it's about a change in a complex system that we still do not fully understand and could include a broad set of possibilities. Wine growing is a long-term commitment, and vintners can plan ahead to face these challenges with solid research and accurate climate modeling. North Fork winemakers have made a living producing outstanding wines in a difficult climate, and I'm sure we will continue. We have always learned to adapt and will undoubtedly do so in the future by experimenting with novel varieties, redefining our terroir, and adapting unique and appropriate wine styles.

Whether one believes climate change is natural or anthropogenic doesn't matter; there's no longer any question that the earth's climate is changing. We must undertake initiatives to try to mitigate human impact and slow the process down to give us more time to adapt to this reality. We're already experiencing earlier starts to our season, hence earlier ripening times. This is nothing new in the wine world, as vintners from around the globe have discussed seeing this same trend for years, especially in the cool climate areas of France, Germany, and Northern Italy. For now, it's a net positive in terms of North Fork wine growing. Longer ripening time can result in higher quality wines. However, time will give us the answers to everything related to growing and making wine in a changing climate.

Predicting climate is a complicated process, but the work of researchers like Dr. Anthony Del Genio and Dr. Ben Cook will help us create new strategies and techniques. We can protect our environment, water, and way of life while continuing to produce some of the most alluring wines in the world. Our best years, I believe, are still ahead.

It Takes a Vineyard

It is easier to build strong children than to repair broken people.

—Frederick Douglass

You may have already noticed that when I talk about wine, I often make comparisons to the human condition. I've found that it's often the best way to help people understand the "mysterious" topic of wine—and sometimes, it helps me to understand wine even better. The similarities are striking.

Take, for example, the life of a vineyard. Grapevines have the same average life span as humans—approximately seventy to eighty years. Initially, young vines can produce a great deal of fruit, yet wine quality may be lacking. Older vines tend to decline in output in their later years, yet they can produce quality fruit and profound, complex wines. Sound familiar?

The issue of terroir becomes much clearer when discussed in sociological terms. Vines, like people, will react to the environmental conditions around them. Both vines and people are biologically the same the world over, yet grown and raised in many different places, taking on the characteristics of where they live. For people, it is speech, customs, and mannerisms; for vines and the wines they produce, it is flavor, aroma, and body. Over very long periods, both will begin to genetically evolve to become more suited to their surroundings. Both will also be affected by outside factors and the influence of others around them. And you thought terroir was just this crazy French concept related to wine.

The concept of early intervention is critical to both vines and children. Like children, vines need to be nurtured early and given adequate care, water, and nutrition to grow well and flourish. Issues caught early on can be mitigated successfully and lead to better outcomes later in life. Grown and raised well, vines and people will last longer and produce better results over the long term. Mismanaged, neglected, and abused, both can produce inferior results and come to a premature end.

The analogy holds when discussing wines in the bottle. Some start out young and attractive—better enjoyed in their youth—only to fade quickly, aging poorly, offering nothing later in life. Others might be fussy and deficient in many ways during their childhood, only to bloom later in life, becoming all that they were expected to be. There are always the ones that start out crummy and end worse. And, of course, we have that segment of the population that truly has their parents to blame. They might have been good grapes when they began, but through lousy winemaking (i.e., parenting), they became rude, obnoxious, and spoiled, providing little enjoyment. The true corps d'elite will always be beautiful and vibrant when young, will age gracefully, and still be attractive and giving in their later years.

There is a belief among some that wine lasts indefinitely—but wine (like people) never stops aging, even when stored under perfect conditions. The culprit that causes aging in both wine and people is the same—oxygen—the double-edged sword of our existence. The myth of immortality has always been a part of human history, but all good things must eventually end. We remember and sing the praise of the good, quickly forget the unremarkable, and forever curse the inferior.

There are many similarities between the human condition and how it relates to wine—where it's raised, how it's grown, and the ways we can guide it, both good and bad. Winemaking—like good parenting—requires a delicate combination of love, intelligent guidance, timing, and discipline. Often a wise, gentle hand, without too much intervention, can be the best combination for success.

No Women, No Wine

If the first woman God ever made was strong enough to turn the world upside down, these women together ought to be able to turn it right again.

—Sojourner Truth

More than ever before, women are in positions of power in industry, law, medicine, and government. In every US presidential election dating back to 1984, women have turned out to vote at higher rates than men. Without the support of women, no candidate of either party can win the US presidency. The advent of the "#MeToo" movement has led to a worldwide awareness of women's rights and their long struggle against exploitation and oppression, the rightful demand of professional equity, and the ability to live free from sexual violence and harassment. This societal reckoning is ongoing and continues to be one of the most important issues of our current age. It has been a long time coming.

The history of women in winemaking goes back thousands of years. The first goddesses of wine were mentioned in stories from ancient Persia and Sumer that predate the Greco-Roman wine gods of Bacchus and Dionysus by thousands of years. Most historians agree that wine was most likely discovered by women, as they were the ones in charge of preparing food. It was women throughout time immemorial who tended the fields, gathered the crops, and in terms of winemaking, harvested the grapes.

Nevertheless, the influence of women in the production and enjoyment of wine typically ended once the grapes were brought in from the fields. Early patriarchal societies defined who could make, trade, and drink wine. In many instances, women were denied the ability to partake in its pleasures. Egyptian hieroglyphic depictions of winemaking rarely show a female form involved in production. Historically, like so many other vocations, the work of winemaking was controlled completely by men. It was only during the height of the Roman Empire (when wine was fully in the male grasp of commerce) that women were formally allowed to drink wine, albeit in regulated moderation.

Over the centuries, women fought for their rights and freedoms under oppressive systems of injustice. From the Middle Ages and through the Renaissance, strong female leaders fought against these restrictions with varying degrees of success. Ultimately, women gained certain levels of parity with men, but the wine industry remained a man's world, longer than many other business sectors. Much of this stemmed from the lack of women's rights to own property until more recent years. Thankfully, the times have changed. So much so that it's become apparent that today's wine industry would be in big trouble without the efforts and contributions of women.

Acclaimed wine writer and critic Jancis Robinson has called the present day the era of "the feminization of wine." Women currently comprise more than half of all US wine consumers; overall, they control $4.3 trillion of the $5.9 trillion annual consumer spending. According to the Wine Market Council, women currently purchase 71 percent of all retail wine sold in America, and annually, about 75 percent of all women will buy wine for their household. This has led to significant marketing efforts toward women consumers. Ultimately, it makes good business sense to ensure that marketing messages appeal to women.[1]

Most importantly, women are fast becoming a significant portion of the industry's leaders. Years ago, winemakers' sons followed their fathers into winemaking while their daughters were groomed for sales or hospitality. Today, women are rising above old stereotypes and gaining significant importance in the wine industry. More women are making their presence felt as oenologists, winemakers, vineyard managers, and cellar workers, while the number of female winery owners, CEOs, and managers continues to rise.

The last twenty years have also seen more women take on the roles of wine directors and sommeliers at some of the nation's top restaurants—positions in the past that were also traditionally held by men. There is now an exclusive wine competition judged solely by women. Known as the *Feminalise*, this tasting event serves as a reference point to guide female buyers and consumers. These developments have also led to changes in the nation's top wine publications. As little as twenty years ago, it was rare for major wine rating organizations to have a female taster on staff. Today, it's clear that their ratings would be far less meaningful without women tasters.[2]

On the North Fork, we have a long and exciting history of women playing crucial roles in our evolution as a region. Louisa Hargrave was a full partner with then-husband Alex in founding the first commercial winery in 1973. Jeanette Smith was the first Cornell University Extension associate, hired in 1984 to work with Suffolk County vineyards. The Cornell Grape Research Program in Suffolk has been led since 1987 by Alice Wise. Women are owners, winemakers, vineyard managers, and sales managers. There are presently at least ten women working on the North Fork as either wine-makers or assistant winemakers with several employed as vineyard managers and assistant vineyard managers.

I've had women working with me in the vineyard and cellar since the early '80s and had harvest crews consisting exclusively of women at various times. Those were some of the best crews I ever had. I was fortunate to be able to work alongside France Posener at The Bridgehampton Winery, who was one of the first women working the wholesale beat in Manhattan; she eventually became the East Coast sales manager for Opus One in Napa. In the last decade alone, I've had more women apply for wine production jobs than men. The tide has clearly turned as women now view wine production and sales as a viable career choice.

Nationally, groups dedicated to women in the wine industry are becoming more common. Deborah Brenner, founder and president of the *Women of the Vine Alliance,* states that "the wine industry is still very much a male-dominated profession. It is not surprising when you delve into wine history and how it evolved around the globe. However, today, women are pushing the boundaries and making amazing strides in all aspects of the wine world. Men and women influence the wine industry, but the approach is slightly different. It is not a gender thing but just reality on how women compromise with each other, share information, seek to collaborate and build relationships."[3]

Julie Brosterman, founder and editor of *Women & Wine,* states, "In every profession, a 'woman's touch' has changed the dynamics, and the wine industry is similar. Men close ranks and have each other's back more than women, who tend to be more independent. They haven't had that 'boys club' benefit that men have had in this industry, but that will happen as there are more and more women in visible roles. The role of women in wine is growing fast. There will be more women in sales than men in the next few years and more daughters taking over wineries as their parents retire. It's fantastic!"[4]

Although women in the industry have made great strides, not everyone believes that true equality has been achieved. Monika Elling, CEO of Foundations Marketing Group in New York City, argues that the imbalance of the sexes goes deep. "Wineries and winemaking are one component where women are more visible, occupying significant roles, but these are not positions that command serious power."[5]

Recent claims of sexual harassment and assault have severely damaged the reputation of the Court of Master Sommeliers, one of the most prestigious organizations in the country bestowing sommelier certification. In an exposé published in the *New York Times* in 2020, several women described the level of abuse they experienced in the male dominant group. "Sexual aggression is a constant for women somms," said one female student. "It's a compromise we shouldn't have to make."[6] It remains to be seen whether subsequent resignations by members of the court and alterations to its policies will lead to any meaningful change. Of the more than 168 people who've achieved the level of "Master Sommelier" from the organization, only twenty-five are women.[7]

For centuries, men have conspired to keep women from achieving positions of influence in the world of wine. Thankfully, we have seen tremendous change as women have begun to break through to the upper echelon of corporate leadership at larger wineries and distribution networks. As with most American businesses, women are gaining influence by the day. I do not doubt that within the coming years, many more women will successfully rise to positions of power in the industry.

The future of the wine industry today is much brighter because of the dedicated work and passion of women. Without the extraordinary contributions of women, it's clear that the North Fork wine region would have never come into existence.

The Local Revolution

I would rather be on my farm than be emperor of the world.

—George Washington

When most people discuss domestic wine, chances are it's from California. More than 90 percent of the wine in the US is made in the Golden State. Most of the remaining 10 percent is produced in the Northwest and East Coast, with the rest spread out over fifty states. It's easy for Californians to enjoy local wine every day. But today, everyone in the nation can enjoy locally made wine and carry the sustainable locavore philosophy into what we drink.

Years ago, eating and drinking locally usually meant walking to the neighborhood pub. Today, of course, being a locavore has taken on a whole new mantra. Much has been written about the farm-to-table movement and local wines. More and more local restaurants are serving foods from farm stands and filling their cellars with wines produced down the road. It takes work, and the chefs that look to their local farmers are recognized and rightly praised for their ideology. Aside from the fact that local wines can be delicious, there are several good reasons to buy local wines. Here is a short list of reasons for drinking locally-produced wine—no matter where you are.

FRESHNESS

Most consumers don't think of the term "fresh" when buying wine, but it applies when talking about local vino. Enjoying a wine made close to home means you get a product that hasn't been jostled and shipped over thousands of miles in an unrefrigerated, over-heated container.

Consequently, local producers can make wines with little or no additives, which are typically needed to ship wine long distances. Local wines are the freshest you will ever taste. Many local wines are released soon after the vintage, still exhibiting a slight "spritz" from fermentation. Then there's the added benefit of being able to sample wine at nearby tasting rooms before purchase, eliminating the guesswork of trying to figure out other people's recommendations. You can visit your local winery, meet the owners and the people working there, and learn a great deal about how they make their wine and what it tastes like before you decide to purchase.

CARBON EMISSIONS AND TRANSPORTATION

A study published in the *Journal of Wine Research* by Tyler Colman and Pablo Päster described their economic model for measuring carbon inputs in a bottle of wine. Their work showed that it's "greener" for New Yorkers to drink wine from Europe with a long sea voyage than wine from California with a long truck trip. This equation held true to the Midwest, the point where wine from Bordeaux and Napa has the same carbon footprint. While reading this study, it was clear that the authors didn't consider that a world-class wine-producing region is less than one hundred miles from the Big Apple. For a New Yorker, drinking wine from the North Fork is the most sustainable choice regarding our carbon footprint. The bottom line is that less fossil fuel is consumed, and less carbon is emitted in getting local wine to market. No other wine is fresher or quicker to market than the one made ninety miles away—especially when you pick it up directly from the winery.[1]

LAND USE

Due to the increase in global demand for wine, new vineyards are being planted all over the world on land that was previously used for "traditional" agriculture or was otherwise in a natural state. Much of California's vineyard land has been converted from prime agricultural land or land once forested. In either case, such change often results in more CO_2 being released into the atmosphere.

On the North Fork, most vineyards were planted on land previously farmed and perilously close to development. Most local vineyards were planted during the land boom days of the early eighties and mid-nineties—both periods of tremendous economic growth. With the advent of farmland preservation programs, the ability for growers to purchase and plant vines significantly increased. In these conditions and within one hundred miles of the nation's largest city, the establishment of the local wine industry was primarily responsible for preventing further development of the North Fork, helping to conserve our precious agriculturally based economy and our sole-source aquifer.

IRRIGATION

The majority of West Coast vineyards are heavily irrigated. The lack of rainfall in the summer and the large water requirement of grapevines makes it nearly impossible to grow grapes commercially in the West without irrigation. Not only can intense irrigation potentially lead to the depletion of local rivers and aquifers, but it also requires energy for pumping. Irrigation water also carries excess agrochemicals into surrounding waterways.

Although many North Fork vineyards have installed drip systems, the size of these vineyards is relatively tiny compared to the vineyards of California. Many of these systems are used only in extreme drought as our natural rainfall, and low yields create a smaller need for artificial irrigation. In most years, we get plenty of rain, providing our vines with all the water they need to grow and prosper.

PRESERVATION OF FARMLAND

In the development pressure of the Northeast, local farming has been losing land to housing for decades. With the advent of the wine industry, the North Fork has preserved over 9,000 acres from encroaching development. Housing developments remove native ground cover and trees, release more carbon, add to water table depletion, and are the primary cause of nitrogen runoff into our surrounding waterways.[2]

ECONOMICS

Buying local wine adds to the regional economy. Local wineries employ neighborhood workers—everyone from field hands to tasting room employees. Large numbers of people are needed to do everything from harvesting grapes and bottling wine to hosting and supporting a wedding reception, providing jobs and opportunities for a wide range of people. Young people working at vineyards and wineries can learn several important skills that can help them later in life. Tasting rooms add substantial revenue to local coffers through property and sales taxes. In the current economy, we need to keep our business local.

REGENERATIVE AGRICULTURE

North Fork vintners have been practicing this form of agriculture for decades before this term became the latest buzzword in farming. The essence of regenerative agriculture is to decrease carbon emissions while increasing carbon uptake and storage. This is accomplished primarily by using cover crops and practicing no-till farming. Permanent cover in the vine rows can sequester up to two tons of carbon per acre, improving soil microbial life and vertebrate biodiversity while increasing water absorption and retention. Local sustainability guidelines from LISW require all growers to have a permanent ground cover between vine rows and apply their composted pomace back into the vineyard. Regenerative agriculture also discourages mechanical tillage under the vine rows, which helps decrease soil erosion and degradation. All these practices can help mitigate the effects of climate change.

VALUE

Local wines are made on a small scale, with comparatively low yields, yet they are far from the most expensive options in the marketplace. There is much diversity in the types of wines available—red, white, rosé, dry, and sweet almost every kind of wine can be found on the North Fork. You can visit local wineries to meet the winemakers and taste what they are offering for sale. Where else can you walk into a place of business and try something before buying? Local wineries allow people to taste and tour to get a feel for the products. North Fork wines are a fantastic value compared to similarly produced, small-batch, hand-harvested wines from other domestic regions.

LOCAL FLAVOR

The term "terroir" is used a lot to describe the flavors of wine. The flavors of grapes grown on the North Fork are unlike any other. Whether you think our wines are better doesn't matter. What does matter is that the aromas and flavors found in North Fork wines are found nowhere else in the world and are worth celebrating.

The French Connection

By three methods we may learn wisdom: First, by reflection, which is noblest; Second, by imitation, which is easiest; and third by experience, which is the bitterest.

—Confucius

The North Fork is blessed with a terroir that can grow most all of the famous grape varieties of Northern and Central Europe. Although vineyard plantings in the '70s and '80s were made up of dozens of different grapes, it was evident early on that some of the most successful varieties grown in North Fork vineyards originated from the French region of Bordeaux. In 1974, two growers planted the first Merlot and Sauvignon Blanc vines on the North Fork—Hargrave Vineyard and Mudd's Vineyard. Many of these vines still exist today and are among the region's oldest. Back then, domestic Merlot was relatively unknown; in 1975, little more than five hundred acres of the variety were planted in the United States, and California had yet to make its mark with the grape. While new entrants to the local wine industry grew lots of different varieties, it soon became apparent that Merlot had the potential to produce some of the best red wines in the region.

The news of Mudd's and Hargrave's success traveled beyond the Fork. Further plantings showed that Merlot had a natural affinity to our local conditions. While a few acres of Cabernet Sauvignon were also in the ground, the wines produced from this king of Bordeaux grapes were less than consistent. Growers saw that Merlot ripened earlier and could thrive in a climate that was often too moist or cool for high-quality Cabernet Sauvignon. The Hargraves discovered that Merlot could produce wonderfully complex red wines with fruit-driven flavors and elegance reminiscent of the

Old World. Many others tasted the results and saw the potential for producing high-quality red wines from this classic grape in our maritime climate.

Mudd's Vineyard was particularly influential in spreading the gospel of Merlot. As a vineyard installation and consultation company, they established over 50 percent of Long Island's vineyard plantings. One of their most consistent recommendations was for planting Merlot. As winemakers developed experience with the variety and saw how well it performed in the cellar, the comparisons to Bordeaux began to gather steam. By the mid-'80s, many wineries released small productions of varietally labeled Merlot to high critical praise. North Fork Merlots were consistently awarded gold medals and "Best of New York" awards in national wine competitions.

Soon after, another classic from Bordeaux arrived on the scene—Cabernet Franc. The first Cabernet Franc went into the ground at Bridgehampton, Bedell, Hargrave, and others. Many planted it to try to faithfully copy the Bordeaux blending scheme for red wines. Vineyard managers discovered that Cabernet Franc grew well in our maritime terroir and could withstand all kinds of conditions. Varietal wines produced from this variety proved to be a revelation, rivaling and often eclipsing wines made from Merlot. The North Fork's ability to consistently produce high-quality red wines from these classic French-Atlantic varieties soon earned it the nickname of "the Bordeaux of the East."

This success led growers and winemakers to drill deeper into the techniques used in the vineyard and cellar. As the East Coast did little research on these grapes and the warmer West Coast climatic conditions were not applicable, we gravitated to the Old World. This led a group of like-minded vintners to host a symposium on the topic in the summer of 1988. The impetus for the conference was primarily influenced by research on the local climate by Cornell University Grape Research Specialist Larry Perrine, which showed the similarities between Bordeaux and the North Fork.[1]

Entitled "Maritime Climate Winegrowing: Bringing Bordeaux to Long Island," the conference hosted an all-star lineup from the famous French region, which included Paul Pontallier, general manager of Château Margaux, Madame May-Eliane de Lencquesaing, proprietor of Château Pichon-Longueville Comtesse de Lalande, and Dr. Alain Carbonneau, director of viticultural research of INRA/Bordeaux, among others. It was a seminal event in the history of our wine district.[2]

The symposium proved a turning point for the industry as North Fork wines were evaluated and compared to our colleagues in Bordeaux. We listened to presentations, tasted wines, and dug holes in vineyards while hanging onto every word spoken by these icons of French wine production. I remember drinking the fantastic lineup of wines they provided over dinner and, later, tasting my 1987 Merlot out of the barrel with Paul Pontallier. Let's just say he was very gracious.

The conference resonated for years and helped producers pay more attention to their vineyards. Our guests implored us to improve vine canopies through leaf removal and coached us on the importance of grape maturity through the expanded use of bird netting to allow more time for fruit ripening. There was some discussion around vine spacing, as the debate surrounding high-density plantings in vineyards was a hot topic at the time. We've since come to understand that, like other aspects of wine production, vine density is just another factor that we should not just mimic. Instead, we've implemented spacings that work best in our region.

The highlight of the conference was not provided by any French visitors but by Larry Perrine. Mr. Perrine's presentation on the climate and soil comparisons of Long Island to Bordeaux turned many heads. It cemented in the minds of the industry what many had assumed—Eastern Long Island's climate was more like Bordeaux than any other wine-growing region in the United States.

In 1990, the symposium concept was revisited—with the emphasis this time on Merlot. The conference star was Michel Rolland, famous for producing Merlot in his native Pomerol and as a consultant to Merlot producers around the globe. More recently, in 2016, LISW sponsored a visit by Dr. Kees van Leeuwen, professor of viticulture at the University of Bordeaux, to assess the region and give a presentation to Long Island growers and winemakers, with an emphasis on the culture and vinification of Cabernet Franc.

I've always believed we could learn more from the Old World than from the West Coast. As part of this philosophy, I also felt that continuing a relationship with Bordeaux winemakers would lead to better wines. In 1997, Stephen Mudd and I reached out to Paul Pontallier to explore a consultation relationship with a new company we had just started working with called Raphael. Paul agreed to work with us, and we began a nine-year consultation partnership. We worked on planning the vineyard, designing the winery, and developing mindful winemaking techniques that suited our terroir. At Paul's urging, we planted other Bordeaux varieties like Malbec and Petit Verdot, which also made delicious wines in our climate and have become mainstays on the North Fork. Working with Paul was a great privilege, and I learned more from him than anyone else in the business.

Direct comparisons of New World regions to the Old World are an age-old problem. On the one hand, it's flattering to be compared to a world-famous wine-producing area. However, just as younger brothers and sisters hate to be compared to their older siblings, we want to be respected for our individuality and take our own path. The reality is that no one region is exactly like another. That's a foundational aspect of terroir. While we learned a great deal from our French cohort, we couldn't lose sight of the fact that we were an East Coast wine region with our own issues and limitations. Over the years, many wineries veered away from focusing on Bordeaux varieties and experimented with other lesser-known grapes like Albariño, Pinot Blanc, Chenin Blanc, and Tocai Friulano, to name just a few. Others delved deeper into Cabernet Franc, Malbec, and Petit Verdot. Some, like myself, did both. While our ability to grow and fully ripen a plethora of European grapes has created delicious diversity in our cellars, our capacity to produce world-class red wines from these famous Bordeaux varieties remains our greatest strength. It is unmatched anywhere on the East Coast.

Raphael Vineyard, Peconic, 1997. Left to Right: Richard Olsen-Harbich, Jack Petrocelli, Paul Pontallier, Beatrice Pontallier, and Stephen Mudd. Author's personal collection.

We learned much about growing and making wine from our friends overseas. That experience can never be replaced and was essential in taking our region to the next level, manifesting itself in the quality of wines we produce today. Since the Bordeaux symposium, we've grown up a lot and learned more about our terroir and what it can make. It's OK to be inspired by another region, but we should not try copying what happens elsewhere. The North Fork is its own place with its own character and conditions. We don't talk much about the comparisons to Bordeaux anymore, but the similarities remain whether we like it or not. I believe that the highest-quality and most age-worthy red wines from our district will continue to be produced from the famous grapes of Bordeaux—but with their own distinct personality and, of course, a certain je ne sais quoi.

Wines of Mass Vinification

To be interested in food but not in food production is clearly absurd.

—Wendell Berry

Believe it or not, there is still some debate going on today that asks if terroir exists. A few books have even been written that refute its existence. I find this astonishing since, to me, terroir is a concept that defines the diversity found in our world. Aren't wines explained and appreciated mainly by where and how they're grown? Isn't that why we have thousands of wines to choose from in stores and restaurants? If terroir didn't exist, what would any region have to offer over any other?

That's before I finally realized how this argument started to develop. It's because of a product that became very popular and produced almost everywhere in the wine world. I call them WMVs—Wines of Mass Vinification.

They're found in any wine shop, or if you live outside New York State, gas station, grocery store, convenience, or state store. WMVs are wines made without any respect for terroir. They are typically produced in large quantities and rarely touch anyone's lips during the winemaking process. Many "wineries" try to make them with the same flavor components for every single vintage. They can do this by utilizing gas chromatography, spectrometry, and flavor profile analysis. Being a winemaker for this type of industrial winery means looking at many charts and lab analyses to determine how your wine will taste.

You can find the ingredients for manufacturing a WMV on the Internet. All the ingredients are legal and approved by the FDA and TTB. Here is just a partial list of them in alphabetical order:

Acacia gum, activated carbon, aluminosilicates, ammonium, ammonium carbonate, calcium carbonate, calcium sulfate, carbohydrase, casein, cellulase, copper sulfate, dimethyl dicarbonate, dimethylpolysiloxane sorbitan monostearate, ethyl maltol, ferrocyanide, ferrous sulfate, fumaric acid, glycerol dioleate, glycerol monooleate, glucose oxidase, hydrogen peroxide, isinglass, maltol, pectinase, phosphate, polyoxyethylene 40 monostearate, potassium bitartrate, potassium citrate, protease, silica gel, silicon dioxide, sorbic acid, soy flour, sulfur dioxide, thiamine hydrochloride.[1]

Some additives are familiar and safe, like pectinase and sulfites. But for the most part, the list indicates that not all wine is made the same. I like my wines to express themselves and emphasize the terroir and the flavor of the region they come from. What if everyone in the world started to look and act the same way, with the same mannerisms, accents, and clothes? It sounds like a bad science-fiction movie.

Finally, there is water. Many West Coast producers rely on water additions to reduce the high sugar content of their grapes grown in an increasingly warming climate. Since Brix levels in grapes rise quickly in hot weather, other constituents like tannins and aromatics can often lag behind. In these cases, wine producers must leave fruit on the vine longer, with ripeness levels approaching as much as 30° Brix, to allow time for these other components to mature. For a winemaker wishing to produce a dry wine under these conditions, excess sugar needs to be reduced. Technology can be used to accomplish this, like spinning cones and reverse osmosis; however, the least expensive way to achieve balance is by adding water. This practice has become common in warm New World areas and has led growers to try to develop different viticultural techniques or investigate planting alternative varieties to combat this issue.

Remember, a Merlot grape (the same one grown in Italy, Sonoma, or Bordeaux) will never make the same exact wine, even if we used the same techniques. In our arrogance, we sometimes forget how little influence we have over the natural world. I want to know what goes into my wine if that's not too much trouble. But when I'm enjoying wine from another region, I want to imagine what that part of the world smells and tastes like and what the people drink. Drinking wine should be like taking a virtual vacation.

Wine should taste good and bring pleasure. And, in a world where everything seems more and more processed and removed from the point of creation, I'd like to think that wine can still be made safe, pure, and free from additives found in many processed foods. If you hadn't noticed, there is no ingredient label on wines. It's too bad because the best ones would have a small, straightforward label. It would look like this:

Ingredients: Grapes, wild yeast.

Wine and Beer

Beer is made by men, wine by God.

—Martin Luther

The craft brewing industry has become hugely successful in the US for a good reason—our beer selections were dominated for years with the same old products, yielding a prosaic and often disappointing array of tastes and aromas. The world of wine is now experiencing what the beer industry went through years ago; consumers want something different and better than the same old, mass-produced libations. They've shown a desire for something local and handmade by passionate people who care about quality and the environment.

Over the last decade, several articles have been written on what craft brewers can teach the wine industry. Winemakers are sometimes criticized for "not taking risks" in their production techniques and for "not being innovative or collaborative" as much as the craft brewing industry. This an interesting opinion, but I beg to differ.

Let's talk first about risk. Wine is grown, not made, and vines can't grow just anywhere. Our main ingredient in wine is grapes, and planting a vineyard requires a huge initial investment and an astute understanding of your location and its potential for fine wine. Of course, there are also the vagaries of weather. As one local vintner once said, "We are in a partnership with Mother Nature, and for better or worse, she is the senior partner." Growing grapes is not for the faint of heart—especially in the Northeast. We have an incredible amount of risk inherent in just bringing our crop to harvest.

Risk-taking is something we manage every day in our cellars. New wine styles and grape varieties are constantly being tested and grown. Wild fermentations—shunned by domestic commercial wineries and academia over the past few decades—have become a more common production method among cutting-edge winemakers. It's not easy to do, and perils abound in the technique, but the results are worth the effort.

While both are fermented beverages utilizing yeast, wine and beer production differ significantly from each other. Brewers can practice their trade and create new products year-round, while winemakers work only with a single annual harvest and have one chance a year to apply their craft. Brewers have the creative freedom and market consent to add flavoring and aromatic additives to make their beers reach new levels of individuality. Consistently producing new and exciting brews is difficult to do without these kinds of additions. Creativity in brewing can come in lots of ways. Some inventive brews include banana, ginger, raspberry, blueberry, apple, cranberry apricot, tangerine, rhubarb, maple, and pumpkin. Some breweries use natural fruits or veggies, though most use an extract, syrup, or processed flavor to create the desired effect. Winemakers that follow these techniques are not taken seriously and are critically panned for not embracing terroir.

Wine's uniqueness comes from the grape varieties and the climate and soil they are grown in. The fruit and herbal aromas and flavors discussed in wines are not added; they are inherently produced in the grapes through the marriage of terroir and fermentation. This doesn't mean that winemakers cannot be more innovative in their approach. The modern marketplace is thirsty for the new and different, and winemakers have responded in recent years by investigating and reimplementing old techniques such as *pétillant naturel*, orange wine, field blending, co-fermenting, and piquette. In each case, the methods are the creative force, producing a wide array of wine styles while respecting terroir.

Wine growers around the world have been collaborating for hundreds of years. On the North Fork, we've only been able to evolve to where we are today through shared research and information. Growing districts have been drawn up and legalized, organizations have been formed for promotion, and research, tastings, and events have been organized for years to showcase the fruits of our labors. More recently, Long Island has developed one of the East Coast's only sustainable certification programs for vineyards, an endeavor that took enormous collaboration between growers who worked closely together to create more eco-friendly practices in local vineyards. These achievements can't happen without a great deal of cooperation and shared vision.

It is good to see craft brewing finally taking a page out of the winemaking notebook by focusing more on the area of origin. Beer brewed through contract brewing and third-party bottlers is no longer a local product. It may not even be the same product originating in the existing facility.

There is still a lot to learn about what makes regional drinks unique, but let's not forget that in the long history of fermentation, wine came first. Without the sweet grapes used to help start the fermentation of malted barley, beer might never have existed at all.

Our Sea-Washed, Sunset Gates

Everywhere, *immigrants* have enriched and strengthened the fabric of American life.

—John F. Kennedy

At the time of this writing, the history of North Fork wine spans almost five decades. From the pioneer days of vineyard and winemaking experimentation to establishing a unique wine style, our region has continued to evolve. We've developed a world-class reputation for fine wine thanks to passionate owners, diligent growers, and creative winemakers who played their part in the development of the district. But something is often left out of this conversation—the significant contribution of the local Latino immigrant population. Working quietly in the background, they don't receive the praise they deserve. It's time to shed some light on those who help make our wine region successful.

Long Island's history of Latino immigration began in the mid-twentieth century and has led to important contributions to the culture and economy of the area. Starting in the 1940s, we've seen three historical waves of Long Island migration. Puerto Ricans made up the first wave of post-war migration to western Long Island. The second wave began with the passage of the Immigration and Nationality Act of 1965, which loosened quotas and opened the door to Dominicans, Ecuadorians, Mexicans, and Colombians, who arrived on Long Island beginning in the 1970s.[1]

The third and most recent wave of Latino immigration started in the late 1980s. It was led by Guatemalans, Salvadorans, and Hondurans fleeing decades of brutal civil wars, natural disasters, and the resulting poverty that ravaged their countries. The most violent phase of Guatemala's thirty-six-year civil war generated significant refugees. Migration continued into the twenty-first century in response to Guatemala's severe socioeconomic problems and increasingly unstable political conditions created from decades of war. This latest wave is most important on the North Fork, where Guatemalan and Salvadoran émigrés find safety and peace working on local farms and vineyards. Migration to Long Island has been, for many, a survival strategy.

Since the 1980s, hundreds of Latino workers have been employed on North Fork farms. Today, local vineyards and wineries employ approximately 250 full-time and 100 part-time Latino workers, accounting for 20 percent of the total agricultural workforce in Suffolk County. They are highly skilled in tending vines—from pruning to shoot thinning to harvest—all critical elements needed to make great wine. These workers have helped take the North Fork wine industry to the next level of success while exhibiting professionalism, dedication, and a passion for producing some of the best wines in the US. These are the unsung heroes of Long Island wine.[2]

It's safe to say that without the Latino migration of the past thirty years, agriculture, as we know it on Long Island, would be a remnant of the past. Our farms, nurseries, greenhouses, and vineyards would have faced severe labor shortages and been unable to sustain the same growth and success. Some would have simply gone out of business. Instead, we're seeing our vineyards thriving and our farmland preserved.

Like the name of our most famous soil, new immigrants from Central America have found a haven in the North Fork wine district. Many of these people are now in vineyard management and winemaking positions at North Fork wineries. We're seeing their children achieve success in our schools and families making a new life in a place where they can live and work peacefully. These people want to give back and create something better—for themselves, their communities, and their families back home, who often have little hope of reaching their dreams.

Great wine is indeed made in the vineyard—and it takes many hardworking, dedicated people in the vineyard to make this a reality. So next time you're enjoying one of your favorite local wines, remember to lift your glass and toast all those who make it possible.

Poseurvores

Hypocrisy can afford to be magnificent in its promises, for never intending to go beyond promise, it costs nothing.

—Edmund Burke

Many of you have seen the famous episode of the television series *Portlandia* where a couple tries to order a local chicken. This spoof about locavores is funny because we all see a little truth in the humor. The farm-to-table movement is one of the most important things to happen to the restaurant world in quite some time. It results from a sustainable philosophy that celebrates and cherishes the local farmer. But as many movements seem to go, it has recently fallen off the rails.

Here's my own version of *Portlandia*. I was recently in a restaurant in Manhattan with my family. The waitress came by to describe the menu selections. She made a big deal out of the fact that many of the vegetables were locally grown, the meat raised Upstate by a friend of the chef, and that the restaurant's overall philosophy was entirely dedicated to the concept of sustainability and supporting local farmers.

While she was talking, I picked up the wine list and began to read through the entries. My mind went blank as I saw one imported wine after another. There was not a single local wine—not one single wine from New York or even the East Coast. I smiled and tried to bite my tongue.

This scene happens often in restaurants across the country, but it's particularly annoying when it's in my hometown. Quite a few restaurants have made a name for themselves and won culinary awards by employing the farm-to-table mantra. I enjoy wines from all around the country and around the world. I'm particularly fond of Old World wines and learn a lot by tasting and drinking them. But if you're going to market yourself as a farm-to-table restaurant, shouldn't you at least try to carry some local wines?

When asked about this issue, restaurateurs and so-called "sommeliers" inevitably get defensive about the subject because they know they've been called out and exposed. Many of these buyers are young and frankly insecure about their wine knowledge and instead rely on the traditional safe choices. It's also true that many distributors and large wineries will "pay to play" to get placed on high-profile wine lists. It's harder for small wineries to play that game, especially when profit margins can be slim to none. For local wineries, it's often tough to be a "hero in your hometown."

I've heard all the excuses. "Local wines aren't good enough, and my customers don't want them." I particularly enjoy the old standby, "they're too expensive," when one look at their wine list would tell you otherwise. And my favorite: "Wine isn't like the other foods. It's not something that can be sold fresh or represent local agriculture. It's more of a commodity." I think wine's preservative nature would add to a restaurant's ability to serve farm-to-table meals year-round. But what do I know? I'm not a restaurateur.

I once heard the following statement from a New York wine writer who agrees with the locavore concept. This writer stated, "Try as America might, stylistically, many wines cannot emulate those of Europe." This old "American wines are heavy with high alcohol" argument completely ignores the existence of an East Coast wine industry. I wonder if these people have ever set foot in a local vineyard.

Let's be clear, growing grapes and making wine is agriculture. We till the soil and grow our crops at the mercy of the weather all year long. Our land is preserved forever as farmland and cannot ever be developed. The vineyards on the North Fork have saved the region from overdevelopment—something that would have surely happened (as it did on western Long Island) had wine producers not arrived in the mid-1970s.

I believe that drinking local wines is an experience one can never have with an imported wine or one that has traveled for weeks and sometimes months to reach its destination. I often bottle wines and have them for sale within the week. Local wines can have a level of freshness that isn't found in the wine you buy off the store shelf. Also, if we're talking about flavor, nothing exemplifies the taste of "local" more than wine. To be clear, winemakers are the folks that essentially created the fundamental concept of terroir.

Farm-to-table cooking is hard to do, especially in the winter. It's never a zero-sum game, as they say. There will always have to be ingredients from other places, whether you're running a restaurant in New York, France, or anywhere else in the world. But one local component is easily accessed all year long. It's local and preserved through the beauty of fermentation, allowing for years of enjoyment. It's a bottle of wine.

If someone is going to market their restaurant with a "farm-to-table" message and advertise the time and energy they put into sourcing locally grown fruit, vegetables, and meat, they're a hypocrite for not carrying local wines. Even better, I have a new name for folks like this—poseurvores.

This is especially true if you are in an area that is near a wine-making region. If you're sourcing local and calling yourself "sustainable," you should also be buying wines that are the same or as close to the concept as possible.

North Fork wines are improving yearly, and many are highly rated. They indeed emulate an Old World style and don't receive the attention they deserve. Wine reflects the taste of where it came from, as well, if not more so, than any plant or animal. In essence, it's a critical part of a local taste experience. To savor the flavors of grass-fed beef and vegetables grown locally and then wash it down with a wine made three thousand miles away makes no sense. If you don't care or don't want to talk about such things in your restaurant, it's a free country, and capitalism is a great system. But when you start making a big deal about being a locavore and creating an image of your establishment as a devotee to local agriculture, you need to put your money where your mouth is.

Are all local wines good? Not at all. No region can make this boast, as you can find poorly made wine worldwide. Can you buy cheaper wines elsewhere? Of course. You can also buy everything else cheaper as well. Most local wines are produced on a small scale; hence, they will not be the same price as large wine brands. The economies of scale is also true for vegetables, grain, fruit, meat, milk, and almost everything else we eat.

Many restaurateurs understand what's at stake. The East End and Manhattan are filled with restaurants that embrace the local food movement. Hundreds of wine lists are full of regional choices, and we're proud of our many relationships with local chefs and wine buyers. Our wineries and restaurants have developed a symbiotic relationship, significantly adding to our area's beauty. It has taken some time to get to this point, and there is more work for us to do as the rest of the greater New York Metropolitan area opens to embrace local wines.

The local food movement came into being not just because local and fresh tastes better but also because the economy surrounding local farms is strengthened and the preservation of farmland is encouraged. Local agriculture improves our lives at the table and through cleaner air and water. This results in a sustainable local economy where revenues are funneled back into the community. What kind of message does it send if a wine list is entirely foreign? I'm sorry, but if a restaurant does this, they are opening themselves up to be guilty of "green-washing" in the least and hypocrisy, pure and simple.

Fashionably Wine

I see that fashion wears out more apparel than the man.

—William Shakespeare

When the first commercial vineyards were planted on the North Fork, many people "in the know" didn't believe our region could successfully grow European wine grape varieties like Chardonnay and Merlot. After all, before Long Island, all the wine produced in New York was in the Upstate districts of the Finger Lakes and the Hudson Valley—places that had trouble growing European varieties in the past and were, at the time, making wine predominantly out of hybrid and native grape varieties. Almost fifty years later, we have repeatedly proved the critics wrong. Today, the North Fork remains one of the world's most innovative and creative wine producers of these two classic varieties. We started producing Chardonnay without oak long before the trend took hold in the rest of the country. The North Fork was also early to the Merlot dance, planting the first vines in 1974. For a time in the early 1980s, Long Island was not far behind California in total acres planted to Merlot.

Merlot is the grape that put North Fork wine on the map and has generated the highest level of critical acclaim from local and national critics. A Bedell Merlot was selected for President Obama's second inaugural luncheon—a profound achievement for our industry. Chardonnay and Merlot remain the highest-scoring Long Island wines in *Wine Spectator*. Whether you agree with the system or not, these varietals have achieved more ninety-point ratings from major publications than any other wines made in our district. Without these two varieties, the face of North Fork wine would be far different and, in my opinion, much less successful. These noble varieties provide a foundation of excellence on the North Fork, whether as single varietals or as the base for creating many of our finest blends.

Any lover of North Fork wines will tell you that the wines we make today are better than what we made in the past. Even with all our success, our wines will continue to improve in quality in the coming years. The reason? Some of it is due to experience, know-how, and an ever-increasing understanding of our terroir. But there's another big reason—clones.

Over the past twenty years, grapevine clones have become increasingly important in our vineyards. Simply put, clones are genetic variants of a particular variety. The Chardonnay grown on Long Island decades ago is not the same vine we have today. Since then, we have benefitted from a wider selection of available plant material. Back in 1980, you only had one choice if you wanted to plant Chardonnay. Today, more than seventy registered clones of this noble white grape are grown worldwide, and they all have their own nuances and characteristics. Many of these clones exist in North Fork vineyards—from the tropical and aromatic *Musqué* to the classic and alluring *Dijon* clones from Burgundy. Although these are all Chardonnays, each exhibits a distinctive character. The success of these clones in our vineyards is directly related to the research trials done by Alice Wise at the Long Island Horticultural Research and Extension Center in Riverhead, NY.

This is also true of grapes like Merlot and Cabernet Franc, where profound differences in wine quality exist between clones grown in the same vineyard on the same soils. Over fifty clones of Merlot are identified in Bordeaux. Pomerol alone has over thirty-five clones of Cabernet Franc. Newer French clones, long kept overseas as tightly held trade secrets, are finding their way into the United States. These new clones are often better suited to our maritime conditions. Some of these clones will ripen earlier, with more maturity than the older selections we used to grow. Some are more resistant to disease. The ultimate result is higher-quality wines. I've seen clones so different from each other that you would think the wines were made from another variety entirely.

All this brings me to my point. The North Fork is sometimes criticized for producing "too much" Chardonnay and Merlot. Some have said these grapes just aren't sexy anymore. Although I don't think one should need a grape to ignite their libido, I believe there is much to say in defense of a stable, long-term varietal relationship. This is especially true when there is so much more to do, many more clone/rootstock combinations to explore, and so many more wines to make. The best is truly yet to come.

Don't get me wrong—many great wines from other varieties are made on the North Fork. It's always exciting to have something fresh and new to the market—especially with a name that no one can pronounce. But let's not forget the partner we first went to the dance with—a pretty good dancer, by the way.

Chardonnay and Merlot remain the most popular and best-selling wines in the world. These grapes are known as classic, noble varieties—a moniker attributed to only a handful of grape varieties. Most importantly, they're delicious and rank among the best (and most expensive) wines in the world when done well. These are the wines that not only helped build the North Fork wine region but also brought fame and fortune to Old World vineyards. In particular, the North Fork remains one of the only places on the East Coast that can grow and ripen Merlot consistently, producing extraordinary wines—something unheard of in this part of the country before we did it. There's something to be said for new and fashionable—but as we all know, there's also something to be said for steadfast dedication and time-honored success.

I think it's important to recognize that from a vineyard and winery perspective, wine fashion is something we need to be careful with. Chasing fads is a common approach for many New World regions. This strategy is lucrative for wineries in the short term but is not a sustainable pathway to maintaining a quality wine district. Chardonnay and Merlot have lost some of their trendiness because of overproduction (and resultant overexposure) in New World areas like the West Coast and Australia. Millions of gallons made from these two grapes have flooded the marketplace over the years—a greedy response to the growing demand. Thankfully, we're not making these kinds of mass-produced wines and have instead gone in the other direction, focusing on genuine wines that are original works, not copies. We encourage low yields, minimal use of oak, elegant extraction, and creative blending that allows our fruit to truly sing a local song.

Of course, there's plenty of room for diversity. Our North Fork climate can support a wide range of varieties from around the globe. I'm incredibly excited about Cabernet Franc, Sauvignon Blanc, and Albariño, which I believe make some of the best wines on the North Fork. We've also had great success with Malbec and Petit Verdot—two varieties that will further improve our reputation for red wines. Other less renowned varieties can be grown successfully on the East End. These can be delicious and fun and give the wine drinker new taste experiences. However, there are good reasons why we don't see more obscure varieties in the greater wine marketplace.

Some are difficult to grow, susceptible to disease, or inconsistent producers. Others make wines that just aren't very interesting.

I would argue that it's not the grape alone that provides a unique wine-drinking experience. The quality that sets North Fork wine apart is not varietal makeup. It's not the fact that we can grow lots of different varieties. On the North Fork, our overall regional style sets us apart. It's about the flavors derived from the surrounding environment and how these are reflected in our wines—no matter what variety or blend. These are wines made with moderate amounts of alcohol—sometimes less than 11 percent, with lots of crisp, juicy acidity and intense aromatics. These wines are not following a fad but are finding their own voice. With our interpretations of the classics, we've helped create a whole new sound. But no matter the variety, our terroir will always provide the loudest instrument in the ensemble.

The world of wine is a fickle place. It can be like the fashion world with styles that come and go. What's hot today is cold tomorrow. It can influence everything from what grapes we grow to how many barrels we should buy. But one thing doesn't change—and that's what grows best in the vineyard.

Take Me Out to the Vineyard

Love is the most important thing in the world, but baseball is pretty good too.

—Yogi Berra

"Is it going to be a good vintage?"

I get asked this question a lot throughout the year. "Are you worried about the weather?" "Is this snow going to affect the wine?" "Are we getting too much rain?" "Is it too dry?" Of course, if you're into wine, these are the questions you like to ask. But the question can never be fully answered for a winemaker until the wine is safely in the tanks. As the great Yankee catcher and Mets manager Yogi Berra once said, "It ain't over till it's over."

The growing season is long—sometimes over 220 days—and lots can happen over the course of seven to eight months. I think some analogies are useful to help folks understand this scenario better. For the avid reader, one might say that a vintage is like a thick, well-written novel that has a large cast of characters and a plot that takes many twists and turns before concluding. For the sports fan, the best analogy to use is the comparison to a baseball season.

Weather is a chaotic system, and as the saying goes, we are in a partnership with Mother Nature, but she is the senior partner. Luckily, our partner is generous, as we are blessed with a climate and soil that can make some of the most distinctive wines in the world. But the fact remains that most of the weather we experience between November and March has little to do with the quality of the upcoming vintage. Mild winters can lead to larger crops, while harsher winter temperatures can decrease yield due to bud damage. But these are quantitative, "off-season" issues. The quality of wines from any vintage is determined between April and October—just like a baseball season. Trying to predict a vintage's quality in April is like asking a baseball fan in the spring who will win the World Series in the fall. One can take a guess, but as they say in sports, that's why you play the game.

Paul Pontallier, the late, great winemaker from Château Margaux, once asked me if I could send him a book about baseball. We were working together in the cellar, making wine at Raphael, and he wanted to learn more about our country. "Why would you want that?" I asked him. He said, "In France, people say that if you want to understand America, you need to learn about the game of baseball." I sent him the best book I could find on the game, but I don't think it made sense to him.

Growing up in the US, most kids are exposed to baseball in one form or another. Almost every child will play it at least once. Mitts, bats, balls, and batting helmets lie around many American children's homes, garages, and apartments. Team logo caps have become ubiquitous. We understand the basics of the game even if many of us have never played or watched it. Many of us know the season's rules, pace, and length. Contrastingly, very few of us grew up with a vineyard in our backyard or a wine cellar in our basement. Paul's idea of learning about America through baseball may not have worked out. However, I believe many Americans can understand more about wine by using the game as a guideline. The analogies between the two fit like a hand in a mitt.

Making wine, like a baseball season, is something that develops over many months. Much behind-the-scenes work occurs during the late fall and winter when the "off-season" tasks are accomplished, and decisions are made to help position vineyards for later success. Like baseball, our season begins in the spring—typically in late April or early May. While the baseball season starts a little earlier, the feeling remains the same; it's a time full of hope and promise—a clean slate when anything is possible. Both seasons get underway when the buds break out, and the first pitch is thrown.

Baseball teams can start the season with a winning streak and remain undefeated for a while before suffering a loss. The opposite is also true; an early losing streak can drop teams far behind in the standings before the schedule is barely a month old. Most of the time, the seasons start the same, with many teams grouped together without a clear picture of how the year will unfold. A baseball team, like a vineyard, will play to its capabilities; the managers and players all have an essential role in its success.

Similarly, a vintage can start slowly or begin hot and fast with lots of growth. We can enter the summer way ahead of what we've experienced in the past, or we can, like many North Fork springs, take some time for summer weather to arrive, needing to catch up during the warmer months to come.

Summer is when the season starts to be more defined as weather patterns develop and persist. Winning days in the summer are sunny, warm, and dry. Losing days are cool and rainy, with loads of humidity that can encourage disease. July and August are the "dog days" of the season, far enough removed from spring and yet too far away from the harvest to be able to gauge what exactly is going to happen. Summer storms and other inclement weather can be overcome and help give each team—and vintage—its identity.

As autumn approaches, things become much clearer. We know a little more about who is winning and who is losing—and who has a chance to make it to the "Fall Classic." The teams doing well are said to be playing "meaningful games" in September. These weeks are crucial to a successful vintage as the weather becomes more tied to success. For both baseball and wine, a great year is determined by what happens in September and October.

Other similarities abound. From the mid-'90s through the mid-2000s, wine, like baseball, went through its "steroid era" when overblown, artificially manipulated powerhouses dominated the American wine scene. Thankfully, the game and the collective psyche of wine drinkers everywhere have turned back to the more traditional pathway, a "balanced approach," if you will, where the natural talent of the players and the fruit are allowed to showcase their collective talents. This is the type of game we play in our region.

Every wine district and vineyard has a unique shape, size, climate, and soil. Successful baseball teams and vineyards are specially built to fit the dynamics of the field they play on. For both ballparks and vineyards, the subtle nuances take time to understand, appreciate, and use to one's maximum benefit. In baseball, this is called the "home-field advantage." In winemaking, we know it as terroir.

Baseball and wine are, at their heart, an escapist pleasure; they're both something people embrace to help relieve the stress of their daily lives. For any baseball team, playing games in October is a winning season. Being on the North Fork means local winemakers can play "meaningful games" every fall. It's what our terroir allows us to do and why we're here in the first place. As I write this, we are still finding out how the current season will turn out, but I'm confident, as I am every season, in our success. However, I think a wise winemaker understands Yogi when he said, "It's tough to make predictions, especially about the future."

But unlike baseball, there are always winners in the game of North Fork winemaking. No matter the weather conditions of the vintage, very good wines can be made every year in our region. In other words, we're set up to achieve sustainable success and can get to the playoffs every year. Some might say this is typical winemaker talk, always saying every year is a good year. I'm here to tell you it's not just spin but the pleasant reality of making wine on the North Fork. And it's true because of our mild temperate climate and just as important—the diversity of the grape varieties planted in our fields.

Now if you don't mind, I need to get back to the cellar and work on getting to the "World Series."

The Wild Oenophiles

All good things are wild and free.

—Henry David Thoreau

The North Fork has long been rich in natural beauty and resources. Before the colonial era, native peoples were hunter-gatherers, feeding off the land that provided an almost unlimited bounty of fish, game, and shellfish. It was only after contact with the early explorers that they developed and fine-tuned their form of local agriculture.

Both native people and colonists supplemented a large part of their diet through hunting and fishing. This natural dynamic continued until the early 1800s, when agriculture became the dominant trade in the region for the next 150 years. After WWII, the demand for more housing and industry began to move from west to east. For a while, it seemed as if Long Island was on the road to complete development as vast tracts of land were eaten up by houses, shopping centers, and blacktop. But agriculture on the East End survived. Today, while the North Fork is home to over two thousand acres of wine grapes, it also remains home to many species of wild fauna.

Growing grapes is not an easy task. Being in the northeast is much more challenging and takes different skills than our colleagues out west. We deal with rain, drought, humidity, cold winters, nor'easters, and hurricanes that can stress vines and create intense disease pressure and injury. Even if you've done your job well and have a clean crop of fruit ripening on the vine, there's always one last obstacle to overcome: wildlife.

Wildlife biologists tell us there are now more white-tailed deer on Eastern Long Island than during the colonial era. The lack of predators has allowed the population to swell; estimates on the North Fork from

Riverhead to Orient Point range between two thousand and five thousand animals—and they all need something to eat.[1]

Deer not only like grapes; they also like almost every part of the vine, from leaves to shoots. They are the most indiscriminate feeders in our vineyards, and growers work hard to discourage them. Deer are gluttons in the vineyard—like an unruly crowd at an all-you-can-eat buffet. They will rip through nets to eat fruit and hurdle through vine rows when chased. Young vines are incredibly delicious to deer, especially the new, tender shoots that first start growing in the spring. Fences are the only way to discourage them.

Rabbits and groundhogs are also fond of tasty young vines, but luckily, they aren't that tall and stay mainly close to the ground. Those blue and tan-colored grow tubes one sees in new vineyards are there to help protect the young vines from being completely devoured by these ground feeders and deer.

Birds are the most significant late-season threat. The North Fork lies within a migratory pathway, and many birds use vineyards as a rest and feeding stop on their way to southern destinations. Unlike deer, they only like the fruit; robins and starlings are the main nuisances, but other birds, such as wild turkeys, also target grapes. Starlings and grackles can form massive flocks and are among the most aggressive eaters, often removing the entire berry. Robins and finches tend to peck at berries, releasing juice and creating opportunities for disease and insects to move in. Left to their own devices, birds will destroy a crop of wine grapes. We really can't blame them. The natural evolution of sweet grapes makes them attractive to birds, which helps the vines reproduce as birds will drop undigested grape seeds to the ground. This is one of the main ways grapevines have spread and developed worldwide.

Years ago, growers tried many different methods to control birds, such as propane cannons and noise machines, making us many friends during the fall. Before widespread bird netting, many growers had to harvest their fruit based on how much damage they would tolerate, often taking fruit before full maturity. With the advent of bird netting, winemakers have the luxury of waiting until the fruit is completely ripe and creating a quieter community. The use of bird netting is one of the main reasons for the increase in wine quality from the North Fork's earlier days.

Raccoons present a whole different problem. Highly adaptable, with no natural predators, raccoons can maneuver under netting and use it as a hammock, lying on their backs while they pick apart grape clusters. They tend not to like the grape skins, so they will pop a berry into their mouths, leaving a tell-tale pile of skins under the trellis.

Luckily not every single animal in our area has a taste for grapes. Fox are carnivores that much prefer to eat grape-fed rabbits, and most birds of prey couldn't care less about fruit. Owls and hawks are mainly carnivores and will feed on those rodents and small mammals that would otherwise dine on grapes, while ospreys prefer fish. The circle of life is on full display in our vineyards.

The "newest" addition to our natural environment is a native species that has made a comeback—the wild turkey. Once common in many regions, overhunting and development led turkeys to virtual extinction on Long Island. None were found from the late 1800s until the late 1900s. Suffolk County and New York State collaborated to reintroduce them in the late 1990s, chiefly in Southaven County Park and Hither Hills State Park. The big birds were right at home, and today we have a population of over six thousand and growing on the East End. So far, they haven't seemed that interested in eating grapes. As turkey populations continue to surge, I hope that remains the case. A welcome habit is their diet of ticks and other insects, along with the seeds of grasses and many weed species.[2] Bald eagles will now occasionally fly over the vineyards as well, so be on the lookout for these magnificent birds as they also make their comeback on the North Fork.

I love seeing local wildlife around the Fork. It's part of our rural character and reflects a vibrant ecosystem. Of course, having a robust wildlife population can also lead to issues around agriculture, but it's all part of farming holistically and is a sign of a healthy environment.

It could be worse; we could be in Tuscany. All they have to worry about are wild boars.

Wine Ratings

As soon as you trust yourself, you will know how to live.

—Johann Wolfgang von Goethe

When it comes to music, cars, sports, food, and politics, Americans are adamant about their likes and dislikes. Why do so many Americans act like "deer in the headlights" when it comes to the subject of wine? How often have you heard someone say, as you pour them a glass of wine, "I'm not a connoisseur!" as if they need to have some special, accrued knowledge? No one says "I'm not a food critic" when eating food or "I'm not a music teacher" when listening to a song they enjoy. Indeed, food and music are two things people understand and are confident about their likes and dislikes; most Americans aren't exposed to wine early, as are many Europeans. The "foreign origin" of wine can create a disconnect for certain people in our country who fight against an elitist image.

Our society is assaulted with lots of misinformation about wine. Social media has no doubt added to the confusion. This and the vast number of wines available to us—especially in New York City—can make the world of wine difficult for the average American to navigate. The truth is that everyone has the tools necessary to judge and appreciate wine—they're called taste buds.

Europeans are raised with at least some understanding of wine from their local region or country. They remember the wines they enjoyed at home or from the vineyard close to where they live. But not all Europeans are educated about the wines of the world; that's something that takes time and experience. Maybe the lesson is that we should start by appreciating our own region's wines before trying to understand the rest of the world.

We've seen the slow changes in American wine consumption due to the well-documented health benefits and the plethora of good wines available at reasonable prices. Still, the average American consumer remains somewhat intimidated by the world of wine. Over half of all wine consumers report wanting to learn more about it. This brings me to the topic of wine ratings.

Much has been written regarding the use of the 100-point scale for wine evaluation. I don't have space in this book to discuss this system's merits and failings, but I would argue that most, if not all, wine drinkers could do fine all by themselves. The problem is that tasting lots of wines takes time and money. As many consumers are unsure of their ability to evaluate wine properly, ratings provide a numeric marker to assist in purchasing decisions.

Most winemakers I know have a love/hate relationship with wine reviews. We love them when we get high scores, validating what we believe our wines to be. We also hate them when we don't get the scores we think our wines deserve. While I believe the North Fork should receive higher ratings from the major publications, the region has been legitimized by the huge number of high-scoring wines over the years. These reviews have led to further respect in the greater marketplace and are useful in helping silence the critics.

Many social media wine pundits are telling us the 100-point scale is fast becoming a relic of decades past when Americans needed the advice of an "expert" to tell them what they should like. The *New York Times* covered the topic by interviewing several significant players in the wine rating world. Most agreed that the 100-point systems used by the *Wine Advocate* and the *Wine Spectator* could be crucial to a wine producer's commercial success. At the same time, however, many of these critics tell us they would like to see the system disappear altogether—that it has become a necessary evil in the wine world. Joshua Greene, the editor, and publisher of *Wine and Spirits* magazine, states that "on many levels, it's nonsensical. I don't think it's a valuable piece of information."[1] This comes partly from the understanding that the 100-point scale is not an exact science. As in all matters concerning taste, it's a matter of opinion. The system used by *Wine Advocate* is different from the one used by *Wine Spectator* and others; the style and type of wine held up to the 100-point standard depend on the taster's preferences. These evaluations are not done completely "blind," leaving much up to the taster's preconceived opinions. Sure, the labels are obscured, but the regional origin of the wines is known to the tasters before appraisement. Add to this the fact that most tasters used by most of the 100-point publications are middle-aged, white males. I can only believe some of these publications are missing a massive part of the equation.

The potential "dark side" of wine ratings was detailed a few years ago when the *New York Times* ran a story about Enologix, a consulting company in California set up expressly to help producers manipulate wines to achieve higher scores. Enologix claims to have supposedly "solved the math of flavor for wine" and figured out how to break down all the components of high-scoring wines so they could be chemically reproduced.[2]

It sounds like a *terroirist's* worst nightmare. Somehow this approach reminds me of the music industry's formula for producing a hit single. Find out what is selling, put together the right faces and voices, and give them a specifically crafted composition to perform. Creative? No, but it does make a lot of money for the people involved. It also begs the question regarding wine and music: Is this all you want to drink or listen to?

The prevalence of social media has led to a new way to spread reviews and opinions. This virtual "word of mouth" (WOM) experience is being led by wine producers, restaurateurs, and sommeliers, along with people who are just passionate about wine. Recent research on the impact of social media has shown that word-of-mouth marketing results in 13 percent of all current business revenue and is currently responsible for over six trillion dollars in annual sales of consumer goods.[3] Indeed, surveys show that 74 percent of consumers identify WOM information as a critical influencer in their purchasing decisions.[4] No doubt, this kind of virtual advertising will continue to grow in importance for small, local wine brands.

Whether it be on Facebook, Instagram, Twitter, or online forums and blogs, wine lovers can now learn more about producers than ever before. Detailed descriptions of winemaking techniques alongside beautiful photography of vineyards and cellars can be found immediately. The result is that consumers can discover and potentially purchase small obscure brands from anywhere in the world.

The benefits of wine scores and reviews can be many. It's helpful to customers and producers trying to navigate the ocean of wine in the marketplace. Local wine skeptics can find hundreds of 90-point scores for North Fork wines going back to the early '80s. Still, real objectivity remains a question with each taster, as so many variables can surround a single review. Professional wine reviewers who are dialed into high alcohol, heavily extracted New World wines will often find North Fork wines lacking in this regard. It shouldn't be held against us; it's simply not what our terroir does.

Let's be honest; wines from more famous regions and producers often benefit from a well-entrenched mythology of supposed "enological greatness" that is rarely challenged by reviewers. Bias in this area is real; I do not doubt that if the wines were truly tasted blind, the North Fork would be receiving a greater number of 90-point scores while more famous producers would endure more criticism. Over the years, many local wineries have held their own or better in private tastings of North Fork wines versus the world's best. But these kinds of comparative tastings would never happen in top wine publications, which are heavily dependent on the big players for advertising dollars. Critics who believe in the mythological version of wine appreciation find it hard to fathom that a relatively new region like the North Fork could ever possibly grow and make wines on a qualitative level to those of the elite Old and New World producers. Many wine professionals said the same thing about California a hundred years ago when they first started. Wine pundits have invested a great deal in creating their own version of reality. Even so, the North Fork is beginning to break through this mindset, which is alarming to many of the old wine guards.

Recently, *Wine Enthusiast Magazine*, a publication that has been rating wine since 1979, announced it would only be reviewing domestic wines made in California, Oregon, Washington, Virginia, and New York, ignoring thousands of good wines produced in forty-five states. This makes WOM marketing even more important for these producers to thrive and succeed. So, while ratings from top publications can be helpful, it's important to remember why wine publications provide ratings; it helps pay the bills. It's best to keep an open mind and allow your taste buds to decide.

We're now seeing customers that are more concerned about *how* wine is made. What kind of vineyard management is employed? Is it sustainable for the surrounding environment and their workers? How and under what conditions were the grapes harvested? Does the wine embody the local character of the region? What, if any, additives are used in the production of the wine in the cellar? And ultimately, does it taste good?

As consumers become increasingly aware of what they put in their bodies, winemaking methods become more important. This often leads consumers to their own local wine regions, where producers are known, techniques are verified, and fruit origin is validated. On the North Fork, our AVA, and the Long Island Sustainable Winegrowing program address these issues and are one of the many reasons these types of programs are so important.

At the end of the day, no one needs to tell you what you should drink. After all, who knows your taste buds better than yourself? The best way to learn about wine is to try lots of different kinds. Local tasting rooms offering samples make this much easier and less expensive. "Practice makes perfect" as they say—and it's much more fun than those piano lessons you had to take. All you need to do is find out what styles of wine you enjoy and, most importantly, learn to trust your palate. As I tell everyone with whom I taste wine, there are no wrong answers regarding your opinion. Forget the background noise and try to have fun. It's not as complicated as you've been led to believe, and your taste buds know much more than you think.

Myth-Busting

Everybody gets so much information all day long that they lose their common sense.

—Gertrude Stein

Being as old as it is, wine appreciation carries many time-honored traditions. Some of these traditions are important and have become widely accepted, such as aging wine bottles on their side or understanding that older vines can produce superior flavors. But misinformation abounds in the wine world. For so long, people were made to feel that they had to have some level of expertise to truly appreciate wine. Ultimately, I believe that wine is made to be enjoyed with as few "rules" as possible, but instances of obfuscation should be clarified. Here is my list of wine facts, which I hope will help resolve some issues.

Fact 1: *You don't need someone to tell you what to drink.*

I don't know about you, but I don't rely on someone else to tell me what I like to eat or what music I should enjoy. Sure, I'll listen to recommendations and read reviews, but ultimately, I'll make my own decisions about what I like. These are personal choices shaped by your experience, environment, upbringing, etc. The same should be true for wine. Sommeliers are good at describing what wines taste like—to themselves. Only you can know what your taste buds are telling you. Don't be afraid to listen to your palate; be confident in your ability to drink what you enjoy.

Fact 2: *There are no wrong answers when describing how a wine tastes.*

Using our senses, especially those of smell, is something we don't consciously do as much as we should. The world around us is moving faster than ever, and we often don't take the time to literally "smell the roses." Our ancestors probably relied on their sense of taste and smell much more than we do for protection, guidance, and safety. You have all the skills at your disposal. Be confident that what you taste and smell is valid. Only you can describe what you are truly experiencing. Embrace it.

Fact 3: *There is no worldwide shortage of cork.*

Despite the increase in the use of screw caps and other alternative closures for fine wine, there is a healthy and sustainable cork industry in the Mediterranean. Natural cork is a renewable resource and part of an agricultural economy covering almost 2.2 million hectares of forested land. Cork harvesting does not harm the tree, and no trees are cut down during the process. Portugal has led the cork industry's sustainable development, implementing important reforestation projects estimated at ten thousand hectares annually. The trade keeps thousands of families employed and provides an essential ecological benefit to the region and the rest of the world. What has changed is the quality of natural cork products. During the past ten years, we've seen a considerable increase in cork quality and a dramatic decrease in the incidence of "corked wines." I expect the science around this issue to get even better as time goes on.[1]

Fact 4: *Sulfites in wine are safe.*

Our country truly has a big misunderstanding about sulfites. For centuries, sulfites have kept wines from turning into vinegar. Sulfur is a certified organic material authorized for use by the National Organic Program. It's never been shown to be harmful to your health, except in the 0.4 percent of the population that suffers from sulfite allergies. In comparison, there are 3.3 million peanut and other tree nut allergy sufferers in the United States. Unlike alcohol, sulfites are not carcinogenic. And sulfites do not cause headaches; if this were the case, people would complain of headaches from dried fruit, nuts, and fruit juice. The headache culprit is probably high alcohol combined with the histamines found in red wine. Once ingested, alcohol is metabolized into acetaldehyde which is proven to cause symptoms of hangovers. Additives in lower-end wines also play a role. The most important thing you should worry about in wine is the alcohol. We know a great deal about that ingredient.

Fact 5: *Organic, biodynamic, and natural wines are not healthier than "conventional" wines.*

Natural, organic, and biodynamic wine adherents continue to claim that wines made this way are healthier for us, with no scientific evidence to back it up. Natural wine ideologues also believe sulfites will hurt you, which is untrue and irresponsible. Grape skins naturally contain anywhere from 6–40 ppm of sulfites. Ranging from 6 to 6,000 ppm, various foods, including fruits and fruit products, vegetables, processed foods, cheeses, and even prescription drugs, pack a much bigger sulfite punch. A two-ounce serving of dried apricots has ten times the level of sulfites of a glass of wine. Research hasn't shown any difference between the effects of biodynamic/natural wine vs. "conventional" wine on the human body—other than they both provide alcohol to your central nervous system.

Fact 6: *Room temperature can't be trusted.*

For years, people were told that red wine should be served at room temperature. That was decent information if you lived in a Scottish castle, but this advice shouldn't be followed today. I'm often served red wines in restaurants at temperatures far too warm. At higher temperatures, red wines can seem overly alcoholic and musky, making them less enjoyable. Red wines taste their best between 55°F–65°F. I like to place red wine in the fridge for an hour or less before serving it to cool it to its proper temperature. They can also be stored in a refrigerator without any issues; take them out an hour before serving. I think you'll find that drinking cooler red wines is way more refreshing and satisfying.

Fact 7: *Climate change is real and is no more apparent than in wine regions.*

The wine industry has been the canary in the coal mine when it comes to understanding climate change. No agricultural sector has older weather records than wine growers. Early religious sects made lots of wine and meticulously recorded harvest dates and weather conditions. We know, for example, that Merlot in Bordeaux was harvested much later in the 1700s than it is today. The same holds true for grape varieties throughout Europe. Winemakers worldwide agree that their fruit is ripening faster than it used to, with harvest dates sometimes two to three weeks earlier than in decades past. Research continues to back up these anecdotal claims. The climate is warming and getting more unpredictable, and wine producers everywhere are working to adjust. If you'd like a good argument, try telling a winemaker that climate change is a hoax.

Tasting Notes

Show me how you drink, and I will tell you who you are.

—Emile Peynaud

Sometimes, it feels like the wine world has become overrun with information. Online wine "experts" are now omnipresent, often providing less-than-accurate information on the topic. Articles abound about "properly" tasting and serving wine, but many are still filled with old advice, perpetuating outdated myths about enjoying wine. As a winemaker, I thought I'd try to clarify what has become a messy discussion. After all, there must be competent wine tasters if you want to make great wines.

First and foremost, wine tasting and drinking wine are very different things. Drinking, of course, is something most of us know how to do. Tasting, on the other hand, is something many people believe they are doing but frequently fail to do properly.

Tasting is how we evaluate and analyze wine to determine quality. Little goes in your mouth, and even less is swallowed. For a winemaker, it's an essential tool that helps guide the creative process. For a consumer, tasting enables you to find out what you like. It's an exercise in perceiving subtleties; the characteristics that make one wine different from another can be minor.

When wine tasting, we look at the color and clarity, smell aromas, and identify flavors. While doing this, little should be imbibed. This is important because the more wine you ingest, the less your taste buds will properly function. Ever heard someone say that the wines they "tasted" while visiting a winery are often better than when they drink them later at home? This is probably because the visitor had already exhausted their taste buds and couldn't accurately judge the wine. Once home and with a fresh palate, the experience can be quite different. Here's the secret: to get the most out of any tasting experience—whether at a winery or home—one must learn how to spit.

I'm often surprised how many winery tasting rooms do not provide a spittoon or "dump bucket." Spitting is an essential part of tasting wine. I know it's not charming, and your mother probably told you not to do it, but if you want instant street cred as a wine connoisseur, start learning how to spit when tasting. By doing so, you'll also train yourself to be able to taste a lot more wines without feeling the effects.

Here are a few more tips I've learned over the years:

Dos:

- Taste in a place free from odors, cleaners, smoke, etc. A room full of cigar smoke, perfume, cologne, or cooking aromas is not conducive to tasting. Think of it as trying to listen to soft music at a low volume. If there is other music playing, outside noise, or people talking to you while you try to listen, it can be very distracting.

- Condition your wine glasses before tasting. Glasses stored in cabinets or china closets often contain residual aromas which can linger. Soap and cabinet/furniture smells can invade an empty glass, affecting a wine's aromas and flavors before you even have a chance to evaluate it. The best way to rectify this is to pour a small amount of wine into the glass (the same wine or something similar), swirl it around, and dump it out. I know it sounds like a waste of wine, but you'll get a more accurate appraisal of what you're tasting.

- What you eat or drink before tasting wine can affect your taste buds. Avoid spicy food, coffee, or tea at least an hour before tasting. The ideal time to taste wine is midway between meals. Paul Pontallier of Château Margaux used to tell me that the best time to taste is "far enough after morning coffee and before lunch"—between ten and eleven o'clock. Of course, that will not work for many people, but you get the idea.

- Swirl the wine around in your glass before smelling it. It's not just a pretentious trick; it allows the wine in the glass to vent off bottle aromas and express its true nature. Swirling briefly raises the level of wine in a glass, covering the sides with a thin layer. Because of the alcohol, this thin layer evaporates, sending aromatics immediately out of the top of the glass. You'll be able to smell the aromatics of wine much more intensely if you swirl it first.

- Sight and smell are also part of the tasting experience. Look at the wine's clarity and color. Is it the color you'd expect that wine to have? Is it clear or cloudy? Does it have an off odor that is unpleasant to you? Your own opinion is all that matters.

- Think of the inside of your mouth as a "globe." Once a little bit of wine is in your mouth, slosh it around and try to get it in contact with all parts of the "globe." The sides of the mouth sense acid, the bottom, sweetness, and the top (the front of the gums) tannin. Many tasters pull air into their mouths, making slurping sounds like when you pull on a straw at the end of a drink. This helps aerate the wine, releasing aromatics as it tumbles around the "globe." Acetic acid (vinegar) is mainly experienced (and felt) as a tingling in your throat.

- Start with younger, dry white wines first, then older, denser, and darker, oak-aged whites. Taste red wines next, starting with the youngest and lightest, ending up with the oldest vintage last. If you taste reds first, the tannins will always make it harder to taste white wines' subtleties and will overemphasize their acidity. Sweet wines of any color should always be tasted at the very end. Tasting sweeter wines first will make dry wines that follow taste bitter.

Don'ts:

- Never pour wine directly into a glass without smelling the empty glass first (see above).

- Don't pour too much wine into a glass for tasting. It's hard to swirl a glass that's too full, and many wines often benefit from more room to aerate. Plus, it's embarrassing to swirl red wine onto your clean white shirt.

- You don't need to smell the cork after it's pulled. A pulled cork can smell bad, while the wine in the bottle can be great. Conversely, a nice-smelling cork can come from a poor bottle of wine. If a server in a restaurant presents a cork to you, just put it on the table and ignore it. The most important thing to do is taste the wine in the glass and bask in the glory that you know more than the person who just handed you a cork.

- Talking about a wine's "legs" is a waste of time. Legs are the drops that run down the inside of a glass after swirling. Wines with large, slow-moving legs have more alcohol, which indicates that they have more alcohol. This is not a quality indicator, and that information can be easily accessed from the label. The best method to determine a wine's quality is by putting it in your mouth and tasting it. A wine containing a high level of alcohol may be unbalanced and taste "hot" (as in boozy), which can undoubtedly be a negative attribute.

Tasting is the most essential tool a winemaker has. It can determine which wines make the cut in a particular blend or, instead, should be combined with another wine to help it become something more. Much of the time, winemakers taste to find flaws. As always, spitting is mandatory. I look for off-odors and flavors, signs that a wine may need to be moved or need some form of intervention. Just as in parenting, you want your children to be healthy and safe but provide them with help and guidance when needed.

Contrary to what many people will lead you to believe, the ability to taste wine is not a God-given trait inherited by a select few. If you can determine what kind of food you like, you can also taste wine. Anyone can learn it through practice and believing in their taste buds. After all, there is no wrong answer to your own experience. Just remember to trust your palate—farther than you can spit.

Our society is often too busy for introspection and reflection. By focusing on things that appear to be important at the moment, we often overlook the many opportunities we have to enjoy and savor the things that can truly bring us joy. Tasting wine is about being mindful and taking the time to appreciate one of the finer things in life. To paraphrase an old saying, slow down, breathe deeply, and take some time to taste the wine.

Wine Futures

It is not in the stars to hold our destiny, but in ourselves.

—William Shakespeare

The spring of 2023 marks the fiftieth anniversary of the modern North Fork wine industry. I don't doubt that many more anniversaries will commemorate the region for years to come. Historically, great wine districts, once developed, don't easily disappear. But as we move into the future, we need to do all we can to nurture the unique gift we've been given and help make it a positive influence on our lives and our community.

It feels like yesterday that I walked into Mudd's Vineyard to begin my career in the wine business. We've come a long way from being a nascent group of growers to the well-established wine region we are today. Our area continues to mature as new investors put down roots. The original industry pioneers are slowly aging out, making way for new ownership and the next generation. Alongside this transition, we retain a core group of established growers and winemakers obsessed with quality, which is evident in their wines.

For many wine-producing regions on the East Coast, a segment of the population remains skeptical regarding quality. Many people across the country (and even many living on Long Island) don't know that the North Fork has some of the best climate and soil in the Northeast and are surprised we can make world-class wines. That skepticism disappears when they taste them. In writing this book, I hope that more people will understand the North Fork and discover the wines we produce.

That said, more people are enjoying North Fork wines than ever before. Over the past decade, consumers have expressed considerable interest in wines with lower alcohol and less extraction. Thankfully, we've moved away from the days when wines were made to please certain critics—high in alcohol, dark as ink, and viscous enough to hold up a straw. This shift in consumer preference has grown out of the locavore and terroir-wine movements. Younger drinkers want their wine to be genuine, honest, and sincerely made, pair well with food, and be crafted with honest, eco-friendly techniques that reflect the terroir in which they're grown. The market is searching for elegance instead of power. Those are the kinds of wine we make—less intoxicating and more refreshing—and our consumers embrace them.

While West Coast producers can still tout the benefits of a warm and dry climate, the precedent for producing high-quality wines in a temperate zone with regular precipitation was set long ago in the Old World. Even in the face of climate change, Old World wines remain lower in alcohol than most New World creations. The growth and acceptance of lower-alcohol wine is a trend that is not going away. Alcohol remains the most dangerous and toxic ingredient in wine and needs to be treated with respect. Plenty of studies have shown the dangers of drinking too much; consuming lower-alcohol wines is clearly better for our health. Lower-alcohol wines also contain significantly fewer calories than higher-calorie, warm-climate wines. Health-conscious drinkers also look for wines that are sustainably produced. This is precisely what the North Fork does.

Our wines have consistently improved over the past few decades and will continue to do so. We are still a relatively young district and producers learn something new with every vintage. The original North Fork plantings were not always the best clones or rootstocks. A significant development has been the understanding and planting of new clone/rootstock combinations. Many of these plantings occurred in the last decade and have already led to dramatic increases in wine quality. We know much more about varieties and clones and where to plant them. This evolution, along with more finesse in the cellar and longer growing seasons influenced by a changing climate, has led to a continual uptick in quality.

As we continue to gain market success and learn more about clones, rootstocks, and ecologically friendly practices, there's no doubt that more vines will go into the ground, and more wineries will be open. But for our wine industry to grow, we'll need to strengthen our research programs, improve consumer outreach and marketing, and demand more collaboration from the local town government. Most importantly, we must continue making the best wines we can.

The growth and expansion of our vineyards are inherently limited by the number of acres available for agriculture and suitable sites for wine grapes. As of this writing, Southold Town has 10,000 acres of farmland out of 34,000 total acres in the township. Over 9,800 acres of land have already been preserved by various levels of government and non-profit organizations through the purchase of development rights and the subdivision process. However, over 45 percent of open land remains unprotected.[1]

Riverhead Town has 17,660 acres of farmland, with 6,893 acres recommended for protection under the Community Preservation Priority Plan. Presently, about four hundred acres of vineyards are planted in Riverhead town. While not every one of these acres is suitable for wine grape production, more than half are, making it theoretically possible to double the size of the North Fork wine industry to approximately five thousand acres. Still, a drop in the bucket when compared to the West Coast.[2]

The biggest issue holding the North Fork back from greater national prominence is size; we are a small area that can't produce enough wine for wider national distribution. I've been asked for years why our wines aren't available in other parts of the country. The truth is, we sell out of everything we make within the greater New York metropolitan area. We could do more national distribution, but our already small profit margins would completely disappear, making the strategy unsustainable in the long run. If more acreage were planted to vines and production expanded, our potential for national exposure would undoubtedly increase. This type of growth is possible but remains to be seen. With land costs increasing due to development pressure, a shortage of affordable housing, an overly regulatory town government, and the increased expense of growing and producing wine in the Northeast, it's unclear whether we will be able to achieve this goal. For now, North Fork wines will remain a regional product.

Since the turn of the last century, comparisons of the North Fork with Bordeaux or the Loire Valley have become passé. Winemakers, consumers, and critics want a narrative built around authenticity, reflecting our local conditions. Due to their small size and dependence on selling wine directly to consumers, North Fork wineries have an inherent tendency to try to produce something for everyone, creating wines that every one of their customers desires. A recent inventory lists at least thirty-one different grape varieties grown on Long Island and more will follow. Unlike our European cohort, we're not subject to AOC-type regulations and can grow whatever grapes we like. During the last two decades, growers and winemakers have expanded their horizons and ushered in a new era of experimentation, reinforcing the concept that diversity is one of our region's greatest strengths. Like all experiments, however, some are successful while others are not.

New varieties are grown for a couple of reasons. One is to embrace the wide array of wine grapes that can thrive here. The other is ongoing; we are exploring varieties that could help us in the future. There's no question that some grapes being considered today wouldn't have been in the conversation years ago. We know more about our climate and soils than ever before and have sharpened our understanding of growing and winemaking techniques. Time is a great educator when it comes to growing wine. We have been doing this long enough to understand what works well and what does not.

A further understanding of our terroir led us to explore white varieties such as Albariño, Viognier, Chenin Blanc, Tocai Friulano, Pinot Gris, Melon de Bourgogne, Grüner Veltliner, Muscat Ottonel, Pinot Blanc, and others. Wines from these grapes have met with great success and critical acclaim. Yet quality wines from red grapes other than the Bordeaux varieties have proved more elusive. Pinot Noir was one of the original varieties planted by the Hargraves and was the first wine they released in 1975. Pinot can make beautiful sparkling wines on the North Fork, but the quality of reds produced from this "heartbreak grape" has been inconsistent over the past fifty years. A handful of dedicated winemakers have occasionally succeeded, catching lightning in the bottle from cooler vintages with less sunshine, but these examples remain rare. Other red grapes such as Syrah, Blaufränkisch, Dornfelder, and Teroldego have also found enthusiastic support among some local winemakers looking to expand their enological horizons and break away from the Bordeaux metaphor. The wines from these red grapes have been intriguing but are mostly underwhelming.

Due to our maritime, lowland geography, common sense dictates that the grapes that thrive along the European Atlantic coast would have the same affinity for parts of the American Atlantic coast. This has proven to be true with the grapes of Bordeaux, the Loire, and Galicia. It will be interesting to see if other grapes at home in European maritime regions will also prove successful on the North Fork. Some of these have already been grown in the Hamptons, including Refosco and Lagrien. Varieties that could also do well are Garganega and Corvina from the Veneto; Romorantin, Folle Blanche, Sacy, Chasselas, and Sauvignon Gris from the Loire; and Mencia, Torrontés, and Loureira Blanca from Northern Spain. A red variant of Melon de Bourgogne is also known to exist in small plantings near Nantes and could be an intriguing choice. Trial plantings of these varieties would be the first step in their evaluation.

Like Sauvignon Blanc in New Zealand, Pinot Noir in Oregon, Riesling in the Finger Lakes, and Merlot on the North Fork, a grape's affinity for a specific area has little to do with comparisons to other places. Numbers and data can show all kinds of relationships. Instead, it's about the confluence of climate, weather, and soil that allows these grapes to feel at home—not a resemblance to their motherland. Putting the "similar climate and soil" narrative aside, the famous grapes of Bordeaux simply like it on the North Fork. But we are not Bordeaux any more than western Oregon is Burgundy, Marlborough is Sancerre, or Seneca Lake is the Mosel. The wine in the glass is all that matters, and it rarely tells a lie.

While the old comparisons with Bordeaux and the Loire may no longer seem relevant given the success of other white varieties, no other European red grapes have yet to excite North Fork winemakers' imaginations like the great reds of ancient Aquitaine and Anjou. Cabernet Franc, Merlot, Malbec, Petit Verdot, and Cabernet Sauvignon, either as varietals or in blends, remain the most compelling red wines in the region. These wines have a juicy, energetic quality on the North Fork, with a flavor and aroma profile that can be complex yet remain fruit-forward and elegantly taut. Ironically, in many years, these wines can be stylistically more like Pinot Noir in Burgundy, with bright acidity, silky tannins, and a level of finesse that can age for decades. This is especially true of North Fork Cabernet Franc; Burgundian winemakers have long characterized this style as "an iron fist in a velvet glove." Yet these wines aren't imitations or copies of some Old World ideal. Instead, they are a reinterpretation of the classics in a unique and regional style that captures every nuance of our terroir. In essence, North Fork wines are both geographically and stylistically, a bridge between the Old and New Worlds of winemaking.

As our climate warms, lengthening our growing season, our wines will undoubtedly improve as we head into the future. No other region on the East Coast has the potential to create such an extensive number of high-quality wines from European varieties. While West Coast areas search for warm-climate varieties to face the oncoming temperature changes, most North Fork growers are content to keep the grapes they already have in the ground and watch them make even better wines. Late-ripening grapes like Cabernet Sauvignon and Petit Verdot will certainly play an essential role in our warming future. Of course, those wine aficionados desiring a big Napa Cult Cab won't be satisfied by what we can do. But those drinkers looking for something new and refreshing in a cool climate style with lower alcohol will gravitate toward our wines.

Our greatest challenge as a region will be our ability to move towards more sustainable practices, find safer materials for disease control, and develop and preserve clean vine material for future plantings, all while navigating the unfolding landscape of climate change. For now, North Fork vineyards are in the vanguard regarding climate resilience, at least for the foreseeable future. While it appears we are currently benefiting from slightly warmer conditions, factors such as extreme precipitation, powerful storms, and the potential for extreme winter cold still exist. Work done by the GISS has indicated that vintages will continue to improve even with increases in precipitation. In the meantime, we must be vigilant and aware of the changing climate as it unfolds around us. Climate change (and what it will bring) needs to be at the forefront of everything we do, from growing grapes to selling wine. We may be experiencing a net benefit from climate change today, but the long-range outlook is less optimistic. We must do everything we can to try to mitigate the effects of climate change by being ready to adapt to dynamic conditions while continuing to improve upon what we already do. It will be challenging to say the least, but I'm confident we will persevere.

As temperatures rise throughout the European Union and the threat of drought and wildfires in the Western US increases, the impact of northern regions like the North Fork on our domestic wine industry will gain importance. Who knows how long Napa Valley will be able to make wines of distinction from the classic noble varieties? Rising temperatures not only create earlier harvests but can lead to a dramatic shift in terroir, resulting in changes in wine style. Extremely warm temperatures can negatively disrupt the balance of sugar, acid, and secondary compounds, changing the rate at which these constituents develop over the growing season. It will become increasingly challenging for the West to produce crisp, refreshing wines without intense manipulation and increasing additions of water and acid. Vineyards growing in drought conditions will require additional irrigation, putting even more stress on water sources. Growers in these areas are already experimenting with warm-climate varieties that may be more suitable to the changing climate. Sustainability will become ever more critical in these areas and others like it around the world.

Likewise, the North Fork and other districts on the East Coast will continue to look for more sustainable ways of wine production. Reducing the use of off-farm inputs like pesticides remains an important goal, particularly when it comes to weed control. Research continues on eco-friendly ways to control competition under the trellis, where it's difficult and labor-intensive to control weeds mechanically. The use of nitrogen and other fertilizers also needs to be minimized due to concerns over groundwater and runoff. Strategies to control new insect pests like the spotted lantern fly and older pests like leafhoppers and mites will need to consider the impact on our ecosystem and our surrounding community. Ultimately, as our climate slowly changes, we need to evolve and be resilient in our strategies for long-term sustainability.

The ultimate long-term solution for growing grapes with minimal off-farm inputs lies in future breeding programs and genetic modification. Breeders in Europe and North America have been developing new grape varieties for decades; disease resistance is the main character trait desired in all programs. Of course, these new varieties must also produce good wines.

A recent decision in 2021 by the European Union granted member states the right to use disease-resistant hybrid varieties with protected denominations of origin (PDO) to help combat climate change. According to the new regulations, member states can now grow both *Vitis vinifera* and hybrid species of *vinifera* crossed with American and Asian vines. Several new hybrid varieties benefit from a higher resistance to downy and powdery mildew, requiring less off-farm inputs and fewer tractor passes through the vineyard, making them more sustainable to grow. These new hybrids have been created using traditional breeding methods of interspecific hybridization and involve several generations of backcrosses and decades of selection. Many are available for commercial propagation in the US; however, significant obstacles remain before these varieties enter the European marketplace as PDO wines. Member states will need to sign off on their use, and the relevant regional authorities ultimately need to approve expanding AVA definitions. Historically, rival breeding programs between France, Italy, and Germany have had problems collaborating, with each country preferring to use and promote its own hybrids. In the meantime, researchers continue to develop grapes that produce desirable characteristics, and growers are experimenting with site selection, growing techniques, and winemaking.[3]

Traditional breeding of grapes can take multiple lifetimes and result in a variety that can be quite different from that of the original plant. New plant breeding technologies (NPBTs) involving genetic modification (GM) have emerged as powerful tools in creating new grape cultivars. Cisgenic plants are similar to traditionally bred plants; cisgenesis involves only genes from the plant itself or a close relative. These are the same genes transferred through traditional breeding, only GM gets the job done in a much shorter time and doesn't alter the genetic heritage of the variety. A disease-resistant, cisgenic Merlot vine would remain varietally Merlot, which would not be possible through traditional or transgenic breeding.

Creating genetically modified clones of centuries-old elite wine grape cultivars remains a polarizing issue; however, the benefits are alluring. What if these grapes could truly be grown without pesticides, fungicides, irrigation, rootstocks, or fertilizer? What if the wine from these grapes was also excellent? The consumer acceptance of cisgenic breeding however, remains to be seen.

Developing grape varieties that would require little to no disease or insect control, minimal inputs of fertilizers, and very little water while producing delicious flavors sounds like an easy decision. This kind of viticulture could make other ecological programs obsolete. It would take decades of research and testing, but the endgame is thrilling to consider.

There's much good that can come out of GM technology. Science itself is neither good nor evil; it's just part of the growing evolution of the human understanding of plant and animal biology. While even rigorously tested GM plants cannot be guaranteed to be 100 percent harmless, not many things that we eat and drink are. The fact remains that any environmental effect of GM plants on our environment will most likely pale in comparison to the human and ecological damage occurring from climate change. Consumers must overcome their fear of GM plants and realize that genetic modification is one of our best tools to combat climate change and environmental degradation. Science and progress got us to where we are today, and we need to rely on science and progress to help lead us into a better future.

The search for indigenous grapes is at the forefront for wine growers worldwide. This trend began years ago in Italy, Spain, and Greece as vintners started to look beyond the commercial appeal of the "noble" grape varieties and embraced their region's unique cultural heritage. It will be interesting to see if the movement back towards indigenous grapes takes hold on the North Fork. For us, this involves the native American grapes shunned by fine wine drinkers for decades.

Although the flavors of native grapes are entirely different, most of them can be grown more sustainably on the East Coast than European grapes. Some exciting choices for lighter reds and eco-friendly viticulture could be old natives like Isabella and Catawba. DNA analysis has determined that these grapes are natural hybrids between a native American vine and an imported European grape that occurred centuries ago, confirming that the early *vinifera* plantings provided some benefits to American viticulture. While these first *vinifera* vines eventually succumbed to phylloxera and mildew, they sowed plenty of colonial oats before they bid the New World adieu. It's possible to grow American grapes organically in the East with minimum chemical inputs for disease control. Although differing in flavor profile from European varieties, these wines reflect a genuine native American personality.

North America is home to more than half of the world's approximately fifty grape species, providing fertile ground for future breeding programs.[4] Again, the question that remains is whether consumers would accept them.

The North Fork gained prominence as a wine region due to its unique ability to grow and ripen the same European varieties that people have enjoyed for centuries. We are one of the few regions in the East that can consistently produce world-class reds from these classic grapes; this accomplishment remains the North Fork AVA's greatest triumph. Since our growing season is favorable to a wide range of European varieties, finding a signature cultivar has been elusive, but it could still arrive in the future. Until then, diversity is our strength.

Today, most North Fork growers feel that planting new varieties to navigate a changing climate is a bit premature—however, we are keeping a close eye on things. In the meantime, the quality of wines from the North Fork will continue to improve, especially with the predicted increases in temperature. It would be a drastic change of direction for the region to focus on native or hybrid grapes. Still, these varieties could play a role in the future of the North Fork wine region as environmental concerns increase along with the search for regional individuality.

One needs to understand the past to gauge North Fork wines' future. So many people are responsible for how far our region has come—from Verrazzano and Hudson, Hargrave, and Mudd, to the youngest wine growers practicing their craft today. I've had the privilege to watch the region evolve from a small handful of growers to one of the country's most diverse and influential wine regions. But the real star of North Fork wine isn't the owners, winemakers, or vineyard managers; it's the terroir itself. Without the blessing we've been given, no amount of talent would make a difference. It's the climate and soil that make it all happen.

Our job now is to make sure we don't screw it up.

Conclusion

Two roads diverged in a wood, and I—I took the one less traveled by, and that has made all the difference.

—Robert Frost

This story is about the North Fork of Long Island wine region. It's a chronicle about passion and the intense desire to succeed against all odds. But it could just as easily be about any local district looking to make wine reflective of terroir.

I'm blessed to have a career I love, one that's given me the ability to raise my family in a beautiful and safe place. I'm also grateful I've been allowed to live and work long enough to share what I've learned. After all, winemaking is about sharing—with friends, loved ones, and our community, along with the thousands of people who've come out to taste North Fork wine. Wine is one of life's pleasures, and I'm happy I've provided that to people. Nothing beats seeing the look of joy on someone's face when they try my wines for the first time. It never gets old.

We as humans often gravitate to wine—in good times and in bad. It helps us celebrate life's accomplishments and milestones while providing comfort, making bad times feel just a little better. It's one of the most beautiful and pleasurable things life offers. It's a big reason why I love making wine.

My late father once told me, "Never tell anyone what you know. Your knowledge is your greatest capital." As a man who witnessed firsthand the horrors of World War II in Europe, I think I knew where he was coming from. He always protected me and implored me to "not give my secrets away" so my job would be secure. I followed his advice for quite some time. However, as I got older, I realized that passing along knowledge was vital for others to succeed. I'm at the point where I don't have any secrets, and I'm comfortable sharing everything I've learned. Hence this book.

The early explorers' predictions of Long Island as a potential wine-producing region have come true. It just took a few hundred years to figure out. As the old American saying goes, "the pioneers take a lot of arrows." The countless attempts at growing European wine grapes helped educate early colonial growers about the limitations of our terroir, and along the way, led to the evolution of unique, native hybrids. By the nineteenth century, it took an international research effort to mitigate the damaging effects of the Columbian Exchange on the global wine industry. Without past failures, we wouldn't have today's successes.

In the years since 1973, when commercial winemaking took hold on Long Island, the United States became a wine-drinking nation. We now buy more wine than any other country. With the influence of social media, we've all become more educated and adventurous about wine. Trends indicate that American consumers are constantly exploring and willing to try new wines and regions. While California has led the way in domestic production, younger consumers are growing tired of the familiar and are looking for new taste experiences. That's very good news for the North Fork.

The North Fork has seen many changes in the past few years. After being "off the radar" for decades, the region has been "discovered" by trend-setters and investors alike. New restaurants opened, and a new generation of farmers took to the land, embracing our mild climate, fertile soils, and proximity to the NYC market. Just when it seemed like everything was going smoothly, the COVID pandemic hit.

Winemaking can't be done remotely. At the onset of the COVID pandemic in March 2020, I never imagined my team and I would spend over a year making wine while wearing masks and trying to socially distance. It was a difficult time for wineries navigating an ever-changing field of regulations and health advisories and even more difficult for employees. I learned how to do virtual meetings and tastings while wondering if we could sell our products. As the Greater NY Metro area shut down, we had to pivot our business model and learn how to sell more of our wine directly to consumers. Our tasting room became a UPS packing area. As of this writing, we're still living with the effects of the pandemic, enduring glass bottle delays, and higher shipping and packaging costs. The war in Ukraine added to supply chain issues as several glass container factories were either damaged or shut down. The labor landscape also changed; people are now less inclined to come to a place of employment and work with their hands.

While existing wages had to increase, the labor shortage continues and may become a lasting effect of the global crisis.

The results of the pandemic weren't all negative. Our ability to ship wine directly to consumers became a large part of our business model as we jettisoned ineffective distributors and put our own sales force on the ground. We also began to infuse more efficient vineyard and cellar practices into our estate, like purchasing a leaf-pulling machine and adding some new equipment to handle fruit at the winery with less labor. The confluence of labor issues and the solutions developed by agricultural technology means more automation is coming in all facets of wine production.

The COVID pandemic helped the North Fork get even more exposure as people traveled closer to home. With over twelve million people living within a hundred-mile radius, we had more visitors to the winery than ever before. As more people began to work remotely, the year-round population of the North Fork increased. People started to drink better wine; many enjoyed North Fork wines for the first time and fell in love with them. Just so, there's a real sense that our industry is changing in many ways, and we need to be able to meet these challenges head-on. The North Fork wine industry has navigated through adversity before, and I have no doubt we will persevere and continue to prove the naysayers wrong.

The fact that the North Fork wine region exists at all is quite an achievement, given our geography and proximity to such large population densities. It is truly a world unto itself, removed from the culture and development of the areas closer to Manhattan. It's an amazingly beautiful place, virtually untouched by modern sprawl, and a national leader in farmland preservation policy. The Long Island Sustainable Winegrowing program became the first on the East Coast, helping educate growers, preserve farmland, and protect our aquifers while carrying our vineyards deeper into ecological balance.

As I try to come up with some final words, I'm getting ready to begin my forty-first harvest on Long Island. Supplies must be ordered, fruit samples collected, and equipment washed and readied. No matter how many vintages I've worked or how difficult the season is, I still get excited for a new vintage. Twenty-twenty-two is shaping up to be a great one, and I can't wait to see how the wines turn out. We had unprecedented heat and hardly any rain to speak of since June—perfect conditions for grapes. I feel the red wines from '22 will be among the best I've ever made and still be drinking nicely long after I'm gone.

It's an exciting time of year—kind of like when the curtain rises and the show begins. Of course, at this point in my life, I feel like an old rocker going out on stage, but it still feels great to make the music of wine. Indeed, my harvest playlists are no longer on cassette tapes or CDs. Now I use Spotify.

Vintages stack up, as do the decades. I can mark different points in time by remembering the weather and conditions of every vintage I've worked. Reflecting on my forty-plus years of tending vines and making wines, I feel a huge sense of accomplishment and peace. I've poured lots of blood, sweat, and tears into my winemaking career, and I'm honored to have been a part of building our region. My children were both born during a harvest, and I've watched them grow into adults, taking enormous pride in them as they build their lives. Most importantly, I'm blessed to be married to the love of my life, who has stayed by my side through every season, every vintage, and every bottle. As I like to say, "happy wife, happy wine!" During the harvest, Nancy was a "wine widow," spending weeks alone raising our children while working full time on her own career as a Child Development Specialist for Cornell Cooperative Extension. I toiled away during harvest, working late hours every day of the week for months. I could not have done any of this without her. I have her insight and expertise to thank for all of my parenting analogies. For all that, and for being my life partner in the vineyard of love we've built together, I am forever grateful.

For in this life, in this beautiful part of the world next to the sea, every day is an extraordinary vintage.

Appendixes

North Fork of Long Island— Original AVA Application (1984)[1]

The following pages show the North Fork AVA application as it was submitted in its original form.

Introduction

This paper aims to provide evidence that the geographical features of the North Fork of Long Island produce viticultural conditions that are distinguishable from the rest of Long Island. The differences are evident in the climate and soil on the North Fork.

The facts presented in the following pages have been accumulated from such respectable sources as Cornell University, the Geneva Agricultural Research Station, the Soil Conservation Service and the U.S.D.A. I feel that the data accumulated in this report clearly shows a need for a separate appellation for the North Fork of Long Island, an area unique within the Eastern United States.

Proposed Appellation

"The North Fork of Long Island"

Origin of the Name

The obvious original use of the term would be the actual way the island "forks" at Riverhead into a North Fork and South Fork. There are several other ways in which the two forks of Long Island have been called:

North Shore/South Shore

North Flukes/South Flukes: thought to be originated by the whaling ships because Long Island looked like a giant fish. The two forks looked like large flukes (fins). Walt Whitman (a native Long Islander) used this term.

North Riding/South Riding: Colonial times. A term for the area covered by colonial judges.

North Line/South Line: Used by the Long Island Railroad.

Geography and Boundaries

The actual geographic area of the North Fork, although attached to a larger island, may be referred to as a peninsula. This is due to the fact that three of its boundaries are surrounded by water: the Long Island Sound to the North, Peconic Bay to the south, and the Atlantic Ocean to the east.

The North Fork region lies entirely in Suffolk County and is governed under the State of New York. As one call see from the U.S. Geological Survey Map (enclosed) the western boundary of the North Fork appellation is the 6-mile-long boundary line separating Riverhead and Brookhaven Townships. The boundary starts at the mouth of Wading River and then becomes a straight-line cutting through Peconic River Park to meet the beginning of the Peconic River. From here on the boundary travels along with the river until it empties into Peconic Bay. It is here that the boundary lines become the three bodies of water. Peconic Bay accounts for the rest of the southern boundary, meeting the Atlantic Ocean at Orient Point. The entire length of the North Fork from its start at Riverhead Town line to Orient

Point is approximately 40 miles. The North Fork is 6 miles wide at its widest point and less than ½ mile at its narrowest. The townships making up the area—Riverhead (78 square miles), Southold (69 square miles) and Shelter Island (11@ square miles)—cover a combined total of 65,000 acres of land or 158 ½ square miles. Shelter Island, although separate from the main strip of land, is composed of the same soil association as those making up the remainder of the North Fork. The towns and villages making up the North Fork include: Aquebogue, Baiting Hollow, Bayview, Blixedon, Calverton, Cedar Beach, Centerville, Cutchogue, Dering Harbor, Eat Cutchogue, East Marion, East Mattituck, Estates of Wading River, Fishers Island, Greenport, Jamesport, Laurel, Mattituck, Nassau Point, Northville, Oregon, Orient, Orient Point, Peconic, Plum Island, am Island, Reeves Park, Riverhead, Roanoke, Robins Island, Shelter Island Heights, Shore Acres, South Jamesport, Southold, Stirling, Wading River, and Waterville.

The total population as of the 1980 was 39,103. The 1984 LILCO estimate is 40,699. This figure also increases greatly when summer residents are counted as well.

There are currently 45 businesses that use the term "North Fork" as part of their name. These companies include banks, churches, real estate companies, etc., and are listed in the enclosed phone listing copy.

History

It is the sea that surrounds Long Island (and more specifically the North Fork) which makes it such a remarkable agricultural area. The "sea renders it more temperate than many other places in the same latitude in the interior." The area is "almost regularly fanned by a breeze from the ocean" and "the air from the sea also has a powerful effect on the climate . . . modulates the heat in the summer and the cold in the winter." It is this moderating effect of the water on the North Fork that makes it an area suitable for fine wine grapes.

When Long Island was discovered by Henry Hudson on September 3rd 1609, he found an island covered with "forests, trees loaded with fruit and grapevines of many kinds." According to Edna Yeager, a historian of the North Fork, the first settlers found that "grapes were just waiting for the winemaker."

The North Fork was the home of several tribes of Indians prior to its settlement by the English. The primary tribes included the Corchaugs. It is from these Indians the original settlers purchased the area known as Southold (the Indians called it "Yennecock") in 1640. This area is roughly equivalent to the boundaries of the proposed viticultural area; the western boundaries of Wading and Peconic Rivers to the eastern boundary of Orient Point.

The English settlers had come from Bingham, England and had settled in New Haven, Connecticut. Due to persecution there, they relocated in Southold. North Fork was considered a part of the New Haven area during this early period. The settlers were very religious—town life was controlled by the church and its leader Reverend John Young. The town of Riverhead was formed from the western part of Southold on 13 March 1792.

The major industry in the area was agriculture. Three hundred and fifty years later it still is. The conservatism of the farmers had helped maintain the area. According to the 1800 census, the North Fork had a population of 12,804. In 1980, 39,103. This represents a small growth when compared to Long Island as a whole. Agriculture was so important to the area that a resident of the North Fork wrote a book on agriculture. Ezra L'Hommedieu of Southold in 1793 wrote "Transactions of the Society for the Promotion of Agriculture, Arts, and Manufacture in New York State."

Past Viticultural History

The settlers trained the native grapes onto arbors behind their homes. Even today, many of the homes have grape arbors. This practice has been going on for as long as the settlers have been here. European wine grapes were not tried on Long Island until the Prince Nurseries started in the late 1700s. One of the earliest viticultural books written in the United States was by William R. Prince in 1830. His *Treatise on the Vine* lists the most favorable soil to grape growing as "light and sandy." This is the soil type of the North Fork. He also states, "light soils . . . when porous, fine, and friable in their composition . . . are the most suitable for the plant and for the quality of the wine."

Prince Nurseries was not on the North Fork (actually it was in Queens), however, they did send grapes to this area. The vinifera (European) grapes did not fare well in the Eastern United States in the 1800s. There was a disease problem that has since been solved.

The backyard arbors were pretty much the extent of grape-growing on the North Fork for most of the next 130 years (from the publication of Prince's book in 1830 to 1963). There were a few attempts at commercial grape-growing but these, too, failed (most notably by a "Moses" Fournier who in the late 1800s planted quite a large vinifera vineyard near Mattituck).

The beginning of the successful commercial vineyards on the North Fork was in March of 1963. It was then that John Wickham, a local fruit farmer, planted a selection of table grapes from Cornell University. So successful was one of the varieties that it was named "Suffolk Red," for the county in New York where it thrived. Mr. Wickham has successfully grown grapes for over 20 years. Prior to his success, vinifera grapes did not survive because of a combination of diseases. The worst on the North Fork (because of the high humidity) was black rot. It is Mr. Wickham's feeling that he succeeded because he used his orchard sprayer in combination with new, more effective fungicides. He stated, "It is the air-blast sprayer that has made grape-growing on the North Fork possible."

The success of John Wickham has led others to the North Fork. It started slowly, but has continued at an accelerated pace the past few years. Professor John Tomkins of Cornell University held conferences in the area in 1968 and 1971. In the "Suffolk County Agricultural News," Volume LV, No. 5, p. 22, he wrote, "there are many good sites for grapes on Long island. Some apple and dairy farmers are taking a real careful look at the opportunities in grape-growing." This was in May 1971.

The June 1971 conferences given by Professor Tomkins were well attended. They were also well reported. Two major newspapers ran articles about grape-growing on the North Fork. It was also Professor Tomkins who steered Alex Hargrave to the North Fork. Hargrave Vineyard was planted in 1973. It was the first commercial vinifera vineyard on the North Fork in the 20th Century. Its success has led to over 1,000 acres of grapes planted in just 12 years. And they are just the beginning.

It has taken over 340 years, from backyard arbors to create a multi-million-dollar industry. But this success was foreseen by many. In the 1800s, Long Island grew many peaches. A Professor Nuttall of Harvard University is quoted by William Prince, "The Peach and the Vine being natural productions of the same region of the East, the opinion has been uniformly adopted, that a climate favorable to the one could not fail to be suitable to the other. And where, let me ask, does the former thrive to a greater degree than in many other sections of our country? From the shores of Long Island . . . the peach flourishes . . . hence we may deduce the surest prospects of an equal success for the vine."

Future Viticultural Outlook

The North Fork is just beginning to break out of its infancy as a viticultural region. Already, it has supported vinifera grapes successfully for over a decade, with the wines produced from this area winning countless awards and praise from critics and consumers alike. As the second decade of North Fork grape growing approaches, much more acreage is expected to produce a full crop as well as new plantings being started. Yet, there is still a tremendous amount of potential for expansion. Hundreds of acres of prime grape land are just waiting to be planted as potato farms and other older agricultural enterprises are finding it harder and harder to survive. The amount of land planted to potatoes is dropping by the thousands of acres annually—the Colorado potato beetle, the diminishing market, and the increasing operating expenses being the main causes of decline.

Currently, there are 5 wineries in operation on the North Fork: Hargrave, Lenz, Jamesport, Pindar, and Peconic Bay Vineyards. On schedule for 1986 are at least 3 other wineries located in Riverhead, Laurel, and Cutchogue. It is very possible that as many as 25–50 wineries could eventually be in operation on the North Fork by the end of this century. With the amount of tourism the area is promoting, and the number of summertime visitors, the North Fork promises great returns on sales and promotion, as well as a good source of labor. Along with these advantages, other assistance is available from both the State and from Suffolk County to the prospective vintner. These include 100% financing and tax abatements, financial assistance on machinery and new building facilities from the Suffolk Industrial Development Agency, New York's Job Development Agency, and the Federal Government's Small Business Administration. Suffolk County's Farm Land Preservation program may also assist the vintner, as well as keeping land available for future vineyard plantings. Also, the North Fork region benefits a great deal from the encouragement of an enthusiastic, progressive local government, who are intensely dedicated to preserving the area's agricultural status.

The North Fork of Long Island and its potential for producing high quality grapes and wine, represents a fantastic opportunity for the prospective vintner. The soil and climate are suited to vinifera grape production like no other area in the East; early results hold great promise for red vinifera varietals such as Cabernet Sauvignon and Merlot. These red varietals have yet to be grown successfully on a commercial scale, elsewhere in the East. Situated only 70 miles from one of the nation's greatest wine markets (New York City) and in the heart of the world's largest consumer market Eastern Seaboard) the North Fork has the potential to become of the greatest wine regions in the United States.

Soils

The grape growing region of the North Fork is encompassed within the area of the towns of Riverhead, Southold, and Shelter Island. This area, when compared to the South Fork, has distinctly different soil types. The difference in soil types begins north of the Peconic River and continues eastward toward Orient Point.

The major soil types that exist on the North Fork, according to the United States Soil Conservation Service, are as follows:

1. Carver-Plymouth-Riverhead Association:

    These soils are excessively well-drained and are very sandy, which may limit its farmability. They are located primarily on the perimeter of the North Fork and are usually rolling or sloping. The natural fertility of these soils is low and the rapid permeability of water through these soils make irrigation a desirable option for vineyards in these areas. They are found mainly along the North Shore adjoining the Long Island Sound.

2. Haven-Riverhead Association:

These soils are characteristically deep and somewhat level and are located further inland on the North Fork. They are well-drained and have a medium texture. Most of these soils have a moderate to high water holding capacity and crops respond well to lime and fertilizer when grown on these soils. Due to these factors, this soil association (which is the predominant one of the North Fork) is considered one of the best farming areas in Suffolk County.

The soils of the South Fork, on the other hand, are somewhat different, and many more associations are present:

1. Plymouth-Carver Association:

These soils are rolling, hilly, deep and excessively drained. Characteristically, scrub oak and other minor trees are found as cover. Permeability is rapid and natural fertility is low. Most of these soils have never been farmed due to these factors and hence they are known to be poor supporters of crops.

2. Bridgehampton-Haven Association:

These soils are deep and excessively drained and have a medium texture. It is its depth, good drainage and moderate to high available water-holding capacity that make this soil well-suited to farming. Most of these areas are currently under cultivation of potatoes and vegetables. These soils are the main reason why South Fork potato and vegetable growers have consistently used less irrigation water than their North Fork counterparts.

3. Montauk, Sandy Variant—Bridgehampton Association:

These soils are deep and usually very sloping. Its steep slopes, irregular topography and a high water table limit the potential of this area for conventional farming, but may be very suitable for supporting grapes. Presently, most of this area is either idle or wooded.

4. Montauk, Sandy Variant—Plymouth Association:

   These soils are excessively drained and coarse textured. Sloping areas within this association also limit conventional farming practices. This loamy-sand is droughty but contains a black surface layer, which is high in organic matter content. There is no indication that grapes cannot be grown on these soils.

5. Montauk-Haven-Riverhead Association:

   These soils are fairly well-drained and are located mainly on the northern side of the South Fork along Peconic Bay. The surface layer is a silt loam, with a fine sandy loam found at deeper levels. These soils are very deep and well-suited to cultivation.

   The remainder of the soils on the South Fork consist of the Dune-Land-Tidal Marsh-Beach Association, which make up the beach and marshland areas, both of which are unsuitable for farming.

Westward into New York City, the soil associations become even more foreign to those found on the Eastern End. It must also be pointed out that while various soil types found in western Long Island may be similar to those found on the North Fork, the encroachment of suburban development and industry on Long Island has made commercial agriculture and land available for it, almost nonexistent in the townships west of Brookhaven (see map).

Soils West of Riverhead

As one can see, the soils of the North Fork and the South Fork are quite different, each giving the grapes that are grown on it a distinct and unique character. At the town of Riverhead where the forks meet, there is still some slight separation of the different soil associations. West of this area, however, the soil associations of Long Island tend to become less restricted to a distinct geographic area and much more intermingling and blending of soil series can be found. Also, there are the soils making up the "engine" of Long Island, namely "The Pine Barrens." The soils of the Pine Barrens can support just that; short, scrubby pine forests are the only vegetation in the light, extremely sandy and infertile soils of this area.

Fortunately, this is the case, as any major agricultural operations or development in the area would limit its ability to be the major ground water recharge basin for Suffolk County (see map). If any small areas were found to be available for grape-growing, however, the light soil would most definitely require some form of irrigation and strict fertilization program if the vines were to survive and be productive.

Land Classes

Land Classes are sub-divisions determined by the SCS to rate the capabilities of various soil series. Most of the soils on the North and South Forks fall into the Land Class members I and II, which state that "the soils contain few or moderate limitations that restrict their use." These are, however, a greater percentage of soil series in the Hamptons which are listed under Land Class III, which states: "these soils have limitations that reduce the choice of plants, require special conservation practices, or both." There is therefore a greater percentage of quality sites available for vineyards on the North Fork.

In general, the soils of the North Fork contain a smaller percentage of silt and loam than the soil series found on the South Fork. This accounts for the fact that South Fork soils have a greater water-holding capacity than North Fork soils and hence require less irrigation. The soils of the North Fork are also generally slightly higher in natural fertility than the soils of the South Fork.

These and other differences which are associated with different soil types and series found on the North Fork can greatly affect the growth of grapes. It is a well-known viticultural fact that particular soils may impart unique balances or combinations of various constituents found within grapes and wines made from those grapes. I therefore feel that the obvious differences in soil types, series, associations, and classes, found between the North and South Fork of the Island as well as between the North Fork and Western Long Island, can impart distinct variations in the components of the grapes and also in the wine made front these areas.

Climate

The climate classification is "humid-continental." However, this is greatly modified by the Atlantic Ocean.

The maritime influence is significant. The surrounding water extends the period of freeze-free temperatures, reduces the range of diurnal and annual temperatures and increases the amount of winter precipitation relative to summer.

Summer Average:	72 F
Winter Average:	33 F
Annual Rainfall:	42" (30" of Snow)
Aug. Wind:	9 mph
Degree Days:	2600–3000

Aug. Frost Free Days: 200–210

Frost Dates

1 yr. in 10	April 30 Spring	Oct. 13 Fall
1 yr. in 5	April 25	Oct. 23
1 yr. in 2	April 14	Nov. 9

Although the North and South Forks of Long Island are relatively close together, there are many climatic differences which exist between these two areas. These differences are due to the unique topography of the Eastern End and the relation of the two forks to the Atlantic Ocean.

Most of the climatic data for the Eastern End of Long Island is recorded mainly from three stations: the Cornell Experiment Station in northeast Riverhead Town, the Greenport weather station, and the U.S. weather station in Bridgehampton. The Cornell Station has been recording weather data since the 1950s, while the Bridgehampton Station has been operating since 1938.

According to this data there are definite climatic differences which exist between the two forks. For example, the average winter temperature on the North Fork is typically 1½ to 2 degrees F. lower than that of the South Fork. This is true even though there are often much lower winter minimum temperatures recorded on the south side for certain cold days of the year. The reason for this is that the North Fork is further away from the Atlantic Ocean and hence does not receive as great an effect from the warmed southwest winds that come in from the Atlantic Ocean. In the winter, the prevailing winds come from the southwest and are warmed slightly by the Atlantic Ocean. In the winter, the sound, bay and ocean have buffering effects due to their accumulation of heat from the summer and fall months.

This wind will therefore buffer the temperature of the South Fork, as it passes over; however, by the time the wind passes over the colder land and Peconic Bay and reaches the North Fork, it has lost some of its warmth and has less of a buffering effect on the temperatures of the North Fork. These breezes, along with those coming off the Long Island Sound, will almost always keep winter minimum temperatures high enough to prevent commercial vine damage.

By the time spring arrives, the ocean has cooled somewhat from the low winter temperatures. Breezes coming from the south at this time of year will therefore become cooled by the ocean, and as they pass over the warming land, a fog will often be produced. This fog will often become trapped on the South Fork and can reduce the accumulation of sunlight and warmth for vine growth. Therefore, in the springtime, the North Fork will usually have more sunshine earlier and also have a higher average temperature. This is evident in the fact that the strawberries, sweet corn and potatoes grown on the North Fork begin to grow and ripen earlier than those same crops grown on the South Fork. Also, the emergence of vine shoots (bud break) is always at least 1 week earlier on the North Fork when compared to the South Fork.

During the summer months the southern breezes coming in off the cool ocean will keep the average temperatures of the South Fork lower than the North Fork. As the winds pass over the South Fork, they travel over Peconic Bay, which is a smaller body of water and hence warmer, increasing the average temperatures of the North Fork. During the summer, the North Fork also receives a greater number of thunder and lightning storms.

These storms usually arrive from the west, and are pushed over towards the North Fork by the prevailing southwest winds.

During the fall, the North Fork of Long Island can also expect slightly warmer temperatures than the South Fork, with the South Fork having cooler nighttime temperatures as well. Otherwise, both Forks have the benefit of enjoying a fall season consisting of a lot of sunshine and normal amounts of precipitation. The ocean effect, which alters the climates of both the North and South Forks is considerably reduced west of Riverhead, where the Island widens. It is this reason along with the increased blending of soil series, which would keep either Fork from being considered part of larger appellation.

Although the amount of sunshine and rainfall can have an effect on the length of the growing season, the single most important factor is the number of days between the spring and fall frosts.

In data taken from the Riverhead station on the North Fork and from the Bridgehampton station, one can see that there are differences in the frost dates for both Forks.

During the 11-year period from 1973–1983, the number of days between frosts, or the length of the growing season is as follows:

Year	Location	# of Days between Frost
1973	Riverhead	207
	Bridgehampton	201
	Greenport	215
1974	Riverhead	194
	Bridgehampton	182
	Greenport	184
1975	Riverhead	192
	Bridgehampton	192
	Greenport	192
1976	Riverhead	197
	Bridgehampton	189
	Greenport	197

Year	Location	# of Days between Frost
1977	Riverhead	216
	Bridgehampton	213
	Greenport	217
1978	Riverhead	197
	Bridgehampton	168
	Greenport	192
1979	Riverhead	200
	Bridgehampton	189
	Greenport	197
1980	Riverhead	196
	Bridgehampton	196
	Greenport	199
1981	Riverhead	176
	Bridgehampton	157
	Greenport	217
1982	Riverhead	178
	Bridgehampton	171
	Greenport	202
1983	Riverhead	190
	Bridgehampton	213
	Greenport	*no data collected

Last Spring Frost

Year	Riverhead	Bridgehampton
1973	April 15	April 21
1974	April 11	April 20
1975	April 25	April 23
1976	April 13	April 13
1977	April 10	April 15
1978	April 4	May 1
1979	April 11	April 22
1980	April 18	April 18
1981	April 21	May 9
1982	April 23	April 23

First Fall Frost

Year	Riverhead	Bridgehampton
1973	Nov. 8	Nov. 8
1974	Oct. 22	Oct. 19
1975	Nov. 1	Nov. 1
1976	Oct. 27	Oct. 19
1977	Nov. 12	Nov. 14
1978	Oct. 18	Oct. 16
1979	Oct. 28	Oct. 28
1980	Oct. 31	Oct. 31
1981	Oct. 14	Oct. 13
1982	Oct. 18	Oct. 11

One can see from this data that in 7 out of the 11 years recorded, there was anywhere from 1 to over 3 weeks longer growing season on the North Fork as compared to the South Fork. This is a very significant difference. When this data is further examined, it was seen that this difference occurs mostly between the dates of the last spring frost. The average last frost on the South Fork is usually around April 23rd, while that on the North Fork occurs around April 14. This spring difference is much greater than the difference between the first fall frosts, which usually occur during the end of October to the beginning of November on both Forks. This supports the fact that the growing season gets off to a quicker start on the North Fork.

The use of heat summation or "Growing-Degree Days" is also another standard for determining climatic differences in grape-growing areas. Heat-summation is a standard developed by the University of California at Davis, and is the measurement of the mean monthly temperatures of a single area, above 50 F. The importance of heat summation above 50 F (10 C) as a factor in grape quality has been indicated by Koblet and Zwicky (1965) and also by Amerine and Winkler (1944). U.C. Davis breaks down various areas in California into 5 climatic regions.

Region

I   Fewer than 2,500 degree days

II   2,501–3,000 degree days

III   3,001–3,500 degree days

IV   3,501–4,000 degree days

V   4,001 or more degree days

The average number of degree days for Riverhead and Bridgehampton area is as follows:

Riverhead (1941–1970)   2,932
Bridgehampton (1941–1970)   2,531

From the period of 1941 through 1970, the average number of heat summation days for the Riverhead station placed them between the Regions I and III. During this same period, Bridgehampton was placed between the Region 1 and 2.

The data for 1973–1979 is as follows:

	Bridgehampton	Riverhead
1973	2714	3200
1974	2392	2800
1975	2734	3131
1976	2457	2925
1977	2692	3100
1978	2382	2750
1979	2652	3100
Average	2572	2987

Once again one can see that during the period of 1973–1979, the area of the Riverhead station on the North Fork varied between Regions II and III while the Bridgehampton area varied between Regions I and II.

As far as grape growing areas are concerned this is a significant difference. In California, many of the appellations are based on the use of heat summation as a cut-off point between two separate growing areas. One can also see from the chart provided, the different areas located within various Regions. These differences can be quite enormous; i.e., Geisenheim, Germany (Region I) and Ramona, San Diego, CA (Region III). In the years 1941–1979, the number of degree days on the South Fork rarely came close to the number accumulated on the North Fork. This is yet another distinguishing climate feature which exists between the North Fork and the South Fork.

Climate West of Riverhead

As the previous data has shown there are quite a few differences between the climate of the North Fork and that of the South Fork.

From the following data, one will be able to see that the climate on the rest of Long Island is also significantly different from the climate found on the North Fork.

Days of Growing Season (days above 32° F) (1973–1982)	
1973	**Growing Season Days**
Riverhead	207
Bridgehampton	201
Brookhaven Lab	137
Patchogue	200
Mineola	234
Central Park NYC	234
1974	
Riverhead	194
Bridgehampton	182

Days of Growing Season (days above 32° F) (1973–1982)	
Brookhaven Lab	149
Patchogue	149
Mineola	192
Central Park NYC	192
1975	
Riverhead	192
Bridgehampton	192
Brookhaven Lab	148
Patchogue	191
Mineola	215
Central Park NYC	204
1976	
Riverhead	197
Bridgehampton	189
Brookhaven Lab	139
Patchogue	163
Mineola	190
Central Park NYC	198
1977	
Riverhead	216
Bridgehampton	213
Brookhaven Lab	156
Patchogue	177
Mineola	216
Central Park NYC	219
1978	
Riverhead	197
Bridgehampton	168
Brookhaven Lab	146
Patchogue	189
Mineola	232

Days of Growing Season (days above 32° F) (1973–1982)	
Central Park NYC	236
1979	
Riverhead	200
Bridgehampton	189
Brookhaven Lab	165
Patchogue	176
Mineola	197
Central Park NYC	236
1980	
Riverhead	196
Bridgehampton	196
Brookhaven Lab	153
Patchogue	188
Mineola	200
Central Park NYC	213
1981	
Riverhead	176
Bridgehampton	157
Brookhaven Lab	155
Patchogue	157
Mineola	224
Central Park NYC	249
1982	
Riverhead	178
Bridgehampton	171
Brookhaven Lab	156
Patchogue	171
Mineola	171
Central Park NYC	222

Last Spring Frost

Year	Riverhead	Brookhaven	Patchogue
1973	April 15	May 8	April 21
1974	April 11	May 8	May 8
1975	April 23	May 8	April 23
1976	April 13	May 13	May 9
1977	April 10	May 4	April 30
1978	April 4	May 4	April 10
1979	April 11	May 3	April 22
1980	April 18	May 10	April 19
1981	April 21	May 9	May 9
1982	April 23	April 30	April 23

First Fall Frost

Year	Riverhead	Brookhaven	Patchogue
1973	Nov. 8	Sept. 22	Nov. 7
1974	Oct. 22	Oct. 4	Oct. 4
1975	Nov. 1	Oct. 3	Oct. 31
1976	Oct. 27	Sept. 29	Oct. 19
1977	Nov. 12	Oct. 7	Oct. 24
1978	Oct. 18	Sept. 27	Oct. 16
1979	Oct. 28	Oct. 15	Oct. 15
1980	Oct. 31	Oct. 10	Oct. 24
1981	Oct. 14	Oct. 11	Oct. 13
1982	Oct. 18	Oct. 3	Oct. 11

The above data shows the differences in growing seasons that can occur, as one moves from eastern to western Long Island. The L.I. Sound, ocean and bay, as described previously, are the main reasons for the North Fork's buffered climate. As the forks merge into the main body of Long Island, the effect of these waters is greatly diminished especially with southwest winds prevailing.

This is evident in the data for Brookhaven and Patchogue, Long Island. Brookhaven, located less than 10 miles west of the North Fork, can have as much as 50 fewer days (almost 2 months) of growing season than Riverhead. Patchogue can also be seen to be as much as 45 fewer days, with most seasons being around 1–2 weeks less than Riverhead. The data given for Mineola (a large suburban area) and Central Park, N.Y.C., show the increasing effect of the buffering ocean winds as the western end of the island begins to narrow once again. A great deal of this effect as well, is most likely due to the great amount of industrial warmth supplied from what is largely an urban area.

The amount of heat summation or "growing degree days" accumulated in areas west of the North Fork also differs considerably. The following data is taken from the Brookhaven National Laboratory.

Growing Degree Days

Year	Riverhead	Brookhaven Lab
1973	3,200	2,560
1974	2,800	2,353
1975	3,131	2,487
1976	2,825	2,299
1977	3,100	2,537
1978	2,750	2,098
1979	3,700	2,486
Avg.	2,987	2,403

Over the period of 1973–1979, Brookhaven averaged 584 fewer growing degree days than Riverhead. This significant difference in heat summation correlates with the shorter growing season found there, as shown previously.

The main reason the climate differs west of the North Fork is due to the smaller effect of the ocean and bay on buffering temperatures.

As the buffering southwest winds approach western Long Island, they first must travel over a small sliver of land known as Long Beach, Jones Beach, and Fire Island (See map).

The winds then must travel over the inlets of South Oyster Bay, Great South Bay, and Moriches Bay, before traveling over the main body of Long Island. The combination of passing over the narrow, colder, island strips and bays causes a slight loss in the warmth of the winds, thereby lessening its effect in buffering the mainland. By the time the winds travel north, a few miles inward over the colder land, they have lost a great deal of the warmth they had previously carried and hence do significantly less to control temperatures, (i.e., frosts) than the breezes traveling over the North Fork.

Conclusion

The data presented in the previous pages clearly shows the unique and distinctive character of the climate and soil of the North Fork of Long Island. The climate and soil have long been understood in viticultural circles as having a very significant effect on the kind and quality of grapes which can be grown in a particular location.

Differences in soil texture have long been studied by the French and German winegrowers as displaying distinctive differences in taste and texture of the wines. There is no doubt that the combination of both climate and soil differences, (along with the examples of wine produced from these two areas in the past) show the viticultural individuality of the North and South Forks of Long Island.

The difference in these two important factors that exists between the North and South Forks of Long Island can have a substantial effect on the growth of vinifera grapes in these two areas. For instance, bud break can occur 1–3 weeks earlier than the emergence of buds on the South Fork. The longer period of growing season encountered on the North Fork favors the cultivation of more later ripening vinifera varieties such as Cabernet Sauvignon, Merlot, and Sauvignon Blanc. These varieties can greatly benefit from a longer post-harvest photosynthetic period, allowing greater strength for winter survival. Late ripening varieties such as Cabernet also need the additional ripening period in order to mature its fruit properly. For these reasons the North Fork is very suitable for red vinifera wine production—something that has yet to be accomplished successfully on a commercial level elsewhere in the East. Also, because of the longer growing season and a greater accumulation of heat units on the North Fork as compared to the South Fork, grapes grown on the North Fork may be able to ripen to a much greater degree and have differing degrees of brix, acidity and pH.

For example, Chardonnay grapes grown on the North Fork may achieve the desired sugar/acid balance acid and/or pH at an earlier date than on the South Fork. Grapes on the North Fork are also growing in soil slightly lighter than those on the South Fork. On some of these soils irrigation may be desirable for the first year or two.

For The Hamptons, Long Island, a different set of climate and soil circumstances leads to other strengths. The cooler temperatures encountered during the growing season on the South Fork can also impart special qualities to wine grapes. Cooler ripening fosters a higher degree of acidity, a lower pH and in some instances may bring to the mature fruit the optimum development of aroma and flavor constituents. Grapes on the South Fork are also growing in soils of a heavier texture requiring less if any irrigation. This factor along with differences on the natural fertility of the soils may also produce subtle differences in the finished wines. Varieties grown in The Hamptons should probably be earlier ripening and highly aromatic in order to benefit from the cooler, shorter growing season.

Along there are distinctive characteristics between the North and South Forks of Long Island, both Forks are clearly different from the rest of the island west of Riverhead town. The reasons for ending the proposed AVAs at the Riverhead/Southampton town-line are numerous. First and foremost, commercial agriculture, and farm land available for its use are quite limited west of the Riverhead area. Running east and west down the center of Long Island are "the Pine Barrens," an untouched pine stand and one of the last wild areas of Long Island. Quite unsuitable for grape production with its extremely light and poor soils, this area is presently being considered by N.Y.S. for preservation status, due to its importance for Long Island's water supply. It seems as though in a few years this area will be off-limits for even recreation, let alone commercial grape production.

The remaining areas available for agriculture, to the north and south of the Pine Barrens, may be suitable for grape-growing, however the differences in both soil and mainly climate distinguish this area significantly from the East End. Apart from various soil types imparting different characteristics, the growing season in this area can be considerably shorter than that found on either Fork. The diminished ocean effect in this area, although is some years similar to the eastern end of Long Island, is very inconsistent, allowing for a greater occurrence of late spring and early fall frosts. The consistently shorter growing season, lower amount of heat summation and lower winter minimums found west of Riverhead greatly increase the threat of winter injury and could force the vintner in this area to carry out cultural practices similar to those used in the colder regions of upstate New York. Certain areas, namely Brookhaven, are probably not even suited to vinifera at all; vinifera grapes need a minimum of 160 days (average) of growing season. This last fact is all the more reason why the western boundary for both AVAs should be the Riverhead and Southampton town lines.

The combination of both soil and climate imparting differences in the constituents of the grapes and wine, the necessity for different cultural practices (i.e., vine-buying, training and spacing) and the possibility of having to grow different varieties (i.e., hybrids, labrusca) reinforces the need for the North Fork appellation to end at the Riverhead town line. It is, therefore, the opinion of these authors, that the proposed boundary for the North Fork appellation defines an area with unique climatic and pedological conditions, different from the rest of Long Island.

The climate and soil of a particular region have been the determining factor for deciding wine-growing appellations in all parts of the globe. The planting of vinifera grapes and the great success achieved in their cultivation on the North Fork of Long Island during the past decade proves that this area has the potential to become one of the finest wine growing areas of the United States.

I therefore feel it is important that the specific grape-growing areas on Long Island be recognized and set apart from one another in order to maintain quality and protect the consumer. The information presented in the previous pages strongly suggests that "The North Fork of Long Island" region has within its boundaries distinct and unique grape growing conditions which warrants the need for approval of a separate viticultural appellation.

AVAs of Long Island

North Fork of Long Island AVA

The famous American poet Walt Whitman like to refer to the East End of Long Island as the North and South Flukes, a term thought to be originated by whaling ship captains because Long Island looked like a giant fish. Early settlers referred to the North Fork by names such as the North Shore or the North Riding, a term for the area covered by colonial judges.

The North Fork of Long Island American Viticultural Area region lies entirely in eastern Suffolk County, Long Island, New York. The western boundary of the North Fork AVA is the 6-mile-long boundary line separating Riverhead and Brookhaven Townships. The border starts at the mouth of Wading River and then becomes a straight-line cutting through Peconic River Park to meet the beginning of the Peconic River. From here on, the boundary travels along the river until it empties into Peconic Bay. It is here that the boundary lines become the three bodies of water. Peconic Bay accounts for the rest of the southern boundary, meeting the Atlantic Ocean at Orient Point.

The entire length of the North Fork from its start at Riverhead Town line to Orient Point is approximately 40 miles. The North Fork is 6 miles wide at its widest point and less than ½ mile at its narrowest. The townships making up the area—Riverhead (78 square miles), Southold (69 square miles), and Shelter Island (11.5 square miles)—cover a combined total of 65,000 acres of land or 158 square miles.

Shelter Island, although separate from the main strip of land, is composed of the same soil association as those making up the remainder of the North Fork. The towns and villages making up the North Fork include: Aquebogue, Baiting Hollow, Bayview, Blixedon, Calverton, Cedar Beach, Centerville,

Cutchogue, Dering Harbor, Eat Cutchogue, East Marion, East Mattituck, Estates of Wading River, Fishers Island, Greenport, Jamesport, Laurel, Mattituck, Nassau Point, Northville, Oregon, Orient, Orient Point, Peconic, Plum Island, am Island, Reeves Park, Riverhead, Roanoke, Robins Island, Shelter Island Heights, Shore Acres, South Jamesport, Southold, Stirling, Wading River, and Waterville.

The 2021 population estimate for Southold Town is 23,836 and Riverhead Town 35,959 for a combined population of approximately 60,000.[2]

The Hamptons, Long Island AVA

"The Hamptons, Long Island" viticultural area is located entirely within eastern Suffolk County, Long Island, New York. The AVA boundaries consist of all the land areas of the South Fork of Long Island, New York, including all the beaches, shorelines, islands, and mainland areas in the Townships of Southampton and East Hampton Gardiners Island, for a total size of 209 square miles. The boundary starts at the intersection of Brookhaven and Southampton Town lines at the Peconic River. It travels south, approximately 10 miles along the Southampton/Brookhaven Township line, until it reaches the dunes on the Atlantic Ocean near Cupsogue Beach in Eastport, NY. Then the boundary proceeds east and west along the beaches, shorelines, islands, and mainland areas of the entire South Fork until it reaches the Peconic River near Calverton at the beginning point. The northern boundary of the AVA is the Peconic Bay, and the southern border is the Atlantic Ocean. Southampton Town has a total area of 139 square miles. East Hampton town consists of 70 square miles and stretches nearly 25 miles from Wainscott in the west to Montauk Point in the east. It is about six miles (10 km) wide at its widest point and less than a mile at its narrowest point. East Hampton has jurisdiction over Gardiners Island, the largest privately owned island in the United States. The 2021 population estimate for Southampton Town is 69,325 and Easthampton Town is 28,512 for a combined population of approximately 98,000.[3]

Long Island AVA

The Long Island American Viticultural Area encompasses Nassau and Suffolk counties of New York, including the smaller offshore islands in those counties. The AVA was established in 2001, over 15 years after two smaller AVAs were created at the eastern end of Long Island. The Long Island AVA designation was promoted as a benefit for some wineries located just outside the two smaller AVAs and for wineries that wanted to create wines that use blends from vineyards in different parts of the Island. It was also developed and promoted as consumer protection of the Long Island name; AVAs require that a minimum of 85% of the fruit used in the designated wine is grown within the region's borders.

Nassau and Suffolk Counties are bounded by the City of New York on the west, Long Island Sound to the north, the Atlantic Ocean to the south, and Block Island Sound and Fishers Island Sound to the east. The western boundary starts at Little Neck Bay at the Queens County–Nassau County border on the north and travels for approximately 16 miles ending at Atlantic Beach on the south. This western border follows exactly the pre-existing border separating Queens and Nassau Counties.

The southern boundary begins from the western point of Atlantic Beach (at the Queens County line). It travels east along the coast following the Atlantic Ocean, encompassing all inlets and barrier beaches, for approximately 100 miles ending at Montauk Point. The northern boundary begins at the Queens County border at Little Neck Bay. It runs east along the island's northern coastline bordering the Long Island Sound, ending approximately 84 miles later at Orient Point. The Long Island AVA area runs in a northeasterly direction and encompasses Fire Island, Robins Island, Shelter Island, Gardiners Island, Plum Island, and Fishers Island.

At Riverhead, 56 miles from the City of New York border, the area separates into two forks, the North and South Forks. The easternmost point on the North Fork is Orient Point, which is 84 miles from the City of New York border, and the easternmost point on the South Fork, is Montauk Point, which is 100 miles from the City of New York border. The width of the area ranges from 12 to 16 miles. There are 1,180 linear miles of shoreline.

Nassau County, which lies to the west of Suffolk County, is 285.4 square miles or 182,680 acres, while Suffolk County is 885.1 square miles or 566,466 acres. Combined, Nassau and Suffolk Counties have 1,170.5 square miles or 749,146 acres.

Nassau County has two cities (City of Glen Cove and City of Long Beach) and three townships (North Hempstead, Hempstead, and Oyster Bay). The towns and villages of Nassau County include Great Neck, Port Washington, Roslyn, Manhasset, New Hyde Park, Mineola, Westbury, Floral Park, Garden City, East Meadow, Elmont, Hempstead, Levittown, Franklin Square, Valley Stream, Wantagh, Lynbrook, Rockville Center, Bellmore, Woodmere, Hewlett, Baldwin, Merrick, Seaford, Inwood, Lawrence, Atlantic Beach, Island Park, Oceanside, Freeport, Bayville, Locust Valley, Sea Cliff, Oyster Bay, East Norwich, Syosset, Greenvale, Jericho, Plainview, Hicksville, Bethpage, Farmingdale and Massapequa.

Suffolk County has ten townships: Babylon, Brookhaven, East Hampton, Huntington, Islip, Riverhead, Shelter Island, Smithtown, Southampton and Southold. The towns and villages of Suffolk County include: Cold Spring Harbor, Huntington, Huntington Station, Dix Hills, Melville, Deer Park, Wyandanch, Lindenhurst, Amityville, Copaigue, Babylon, Centerport, Northport, Kings Park, Commack, St. James, Smithtown, Hauppauge, Brentwood, Bay Shore, Brightwaters, West Islip, Islip, Great River, Oakdale, Bayport, Sayville, Bohemia, Central Islip, Holbrook, Setauket, Stony Brook, Centereach, Lake Grove, Lake Ronkonkoma, Holtsville, Medford, Patchogue, Bellport, Brookhaven, Yaphank, Farmingville, Selden, Coram, Middle Island, Mount Sinai, Port Jefferson, Miller Place, Rocky Point, Shoreham, Ridge, Shirley, Mastic, Mastic Beach, Center Moriches, Calverton, Wading River, Roanoke, Eastport, Westhampton, Riverhead, Northville, Flanders, Quogue, Hampton Bays, Southampton, Watermill, North Sea, Noyack, Bridgehampton, Sag Harbor, East Hampton, Amagansett, Napeague, Montauk, North Haven, Orient, East Marion, Greenport, Southold, Peconic, Mattituck, Cutchogue, Jamesport and Aquebogue.

The 2021 population estimate for Nassau County is 1,390,907, for Suffolk County 1,526,344 for a combined population of 2,9217,251.[4]

North Fork Growing Degree Days 1988–2022

North Fork Growing Degree Days			
1988	3191	2010	3762
1989	3314	2011	3544
1990	3366	2012	3614
1991	3692	2013	3300
1992	2859	2014	3264
1993	3190	2015	3511
1994	3177	2016	3539
1995	3194	2017	3436
1996	3035	2018	3580
1997	3176	2019	3458
1998	3543	2020	3467
1999	3575	2021	3651
2000	3213	2022	3509
2001	3370		
2002	3468		
2003	3096		
2004	3287		
2005	3452		
2006	3244		
2007	3494		
2008	3201		
2009	3024		

Notes

Note to the Introduction

1. Carl L. Becker, "Everyman His Own Historian," *American Historical Association (AHA)*, accessed 23, 2022, https://www.historians.org/about-aha-and-membership/aha-history-and-archives/presidential-addresses/carl-l-becker.

Notes to History

1. "The Discovery of New York," *Castello Di Verrazzano: Greve in Chianti: Italy*, accessed 2022, https://www.verrazzano.com/en/la-scoperta-di-new-york/.

2. "A Tour of New Netherland: Long Island," *New Netherland Institute*, accessed November 23, 2022, https://www.newnetherlandinstitute.org/history-and-heritage/digital-exhibitions/a-tour-of-new-netherland/long-island/.

3. Adriaen van der Donck, Ch. T. Gehring, and Diederik Willem Goedhuys, *A Description of New Netherland* (Lincoln: University of Nebraska Press, 2010).

4. Crystal A. Dozier, Doyong Kim, and David H. Russell, "Chemical Residue Evidence in Leon Plain Pottery from the Toyah Phase (1300–1650 CE) in the American Southern Plains," *Journal of Archaeological Science: Reports* 32 (2020): 102450, https://doi.org/10.1016/j.jasrep.2020.102450.

5. Richard Figiel, *Circle of Vines* (State University of New York Press, 2014), accessed 30, 2022, https://sunypress.edu/Books/C/Circle-of-Vines.

6. Figiel, 2014.

7. Figiel, 2014.

8. William Robert Prince and William Prince, *A Treatise on the Vine: Embracing Its History from the Earliest Ages to the Present Day, with Descriptions of above Two Hundred Foreign and Eighty American Varieties: Together with a Complete Dissertation on the Establishment, Culture, and Management of Vineyards . . .* (New York: T. & J. Swords, 1830).

9. Prince and Prince, 1830.

10. Prince and Prince, 1830.

11. William Robert Prince and William Prince, *A Treatise on the Vine: Embracing Its History from the Earliest Ages to the Present Day, with Descriptions of above Two Hundred Foreign and Eighty American Varieties: Together with a Complete Dissertation on the Establishment, Culture, and Management of Vineyards* . . . (New York: T. & J. Swords, 1830).

12. Richard Figiel, *Circle of Vines* (State University of New York Press, 2014), accessed 30, 2022, https://sunypress.edu/Books/C/Circle-of-Vines.

13. "Cooperative Extension Association of Suffolk County," *Suffolk County Farm and Home Bureau News*, volume 55, Google Books (Cooperative Extension Association of Suffolk County), accessed September 23, 2022, https://books.google.com/books/about/Suffolk_County_Agricultural_News.

Notes to The Land

1. *Garvies Point Museum and Preserve*, "Geology of Long Island," accessed December 30, 2022, https://www.garviespointmuseum.com/geology.php.

2. "Long Island: Our Story / Chapters 1–5," *Newsday*, February 27, 2019, https://www.newsday.com/long-island/long-island-our-story-w53584.

3. Leslie A. Sirkin, *Eastern Long Island Geology: History, Processes, and Field Trips* (Watch Hill, RI: Book and Tackle Shop, 1995).

4. Sirkin, 1995.

5. Sirkin, 1995.

6. "Long Island: Our Story / Chapters 1–5," *Newsday*, February 27, 2019, https://www.newsday.com/long-island/long-island-our-story-w53584.

Notes to The Soil

1. P. V. Krasil'nikov et al., *A Handbook of Soil Terminology, Correlation and Classification* (London: Routledge, 2016).

2. John W. Warner, "Soils of Suffolk County, New York," Washington: US Government Print Off, 1975.

Notes to The Sea

1. Wikipedia, s.v. "Long Island Sound," last modified February 24, 2022, https://en.wikipedia.org/wiki/Long_Island_Sound.

2. *SeaTemperature.info*, "Long Island Sound Ocean Water Temperature," accessed December 3, 2022, https://seatemperature.info/long-island-sound-water-temperature.html.

3. *SeaTemperature.info*, "Peconic Bay Ocean Water Temperature," accessed August 30, 2022, https://seatemperature.info/peconic-bay-water-temperature.html.

Notes to The Sun

1. Krysten Massa, "Is Cutchogue Really the Sunniest Spot in New York?" *Suffolk Times*, June 4, 2016, https://suffolktimes.timesreview.com/2016/06/is-cutchogue-really-the-sunniest-spot-in-new-york/.

2. Aileen Jacobson, "A Summer Place, Magnetic Year-Round," *New York Times*, May 4, 2012, https://www.nytimes.com/2012/05/06/realestate/cutchogue-li-living-in-a-summer-place-magnetic-year-round.html.

3. Krysten Massa, "Is Cutchogue Really the Sunniest Spot in New York?" *Suffolk Times*, June 4, 2016, https://suffolktimes.timesreview.com/2016/06/is-cutchogue-really-the-sunniest-spot-in-new-york/.

4. *EPA (Environmental Protection Agency)*, "Sun Safety Monthly Average UV Index," accessed December 30, 2022, https://www.epa.gov/sunsafety/sun-safety-monthly-average-uv-index#tab-9.

5. Ibid.

6. Ibid.

7. Ibid.

Notes to the Aquifer

1. *Nassau Suffolk Water Commissioners' Association*, accessed July 7, 2022, https://www.nswcawater.org/water_facts/our-long-island-aquifers-the-basics.

2. *Cornell Cooperative Extension*, "Precipitation Reports 2020–2029," accessed July 6, 2022, https://ccesuffolk.org/agriculture/growing-degree-days-gdd/gdd-from-previous-years/precipitation-reports-2020-2029.

3. Ibid.

4. *Department of Health*, "The Suffolk County Comprehensive Water Resources Management Plan Team, Suffolk County Department of Health Services," accessed July 3, 2022, https://www.health.ny.gov/prevention/public_health_works/honor_roll/2016/suffolk_cwrmpt.htm.

5. *Suffolk County Government*, "Comprehensive Water Resources Management Plan," accessed March 27, 2023, https://www.suffolkcountyny.gov/Departments/Health-Services/Environmental-Quality/Water-Resources-Management-Plan.

Notes to Climate and Weather

1. *CLIMOD 2*, accessed December 30, 2022, http://climod2.nrcc.cornell.edu/.
2. *WillyWeather*, "North Fork Wind Forecast," accessed 2022, https://wind.willyweather.com/ny/suffolk-county/north-fork.html.
3. *Cornell Cooperative Extension*, "Growing Degree Days (GDD)," accessed December 2, 2022, https://ccesuffolk.org/agriculture/growing-degree-days-gdd.
4. Ibid.
5. *CLIMOD 2*, accessed July 7, 2022, http://climod2.nrcc.cornell.edu/.
6. Ibid.

Notes to Growing Degree Days

1. M. A. Amerine and A. J. Winkler, "Composition and Quality of Musts and Wines of California Grapes," *Hilgardia* 15, no. 6 (February 1944): 493–675, https://doi.org/10.3733/hilg.v15n06p493.
2. A. J. Winkler et al., *General Viticulture*, 2nd ed. (Berkeley: University of California Press, 1974).
3. *Vanity Fair*, "An American Original," October 6, 2010, https://www.vanityfair.com/news/2010/11/moynihan-letters-201011.
4. *Best Places*, "Climate in Westchester County, New York," accessed 2022, https://www.bestplaces.net/climate/county/new_york/westchester.

Notes to The North Fork vs. the Hamptons

1. *Cornell Cooperative Extension*, "Precipitation Reports 2020–2029," accessed November 3, 2022, https://ccesuffolk.org/agriculture/growing-degree-days-gdd/gdd-from-previous-years/precipitation-reports-2020-2029.
2. *Cornell Cooperative Extension*, "Growing Degree Days (GDD)," accessed November 2, 2022, https://ccesuffolk.org/agriculture/growing-degree-days-gdd.
3. Ibid.

Notes to Terroir

1. Madeline Puckette, "Looking for Good Wine? Start with the Appellation," *Wine Folly*, accessed January 5, 2022, https://winefolly.com/deep-dive/looking-for-good-wine-start-with-the-appellation/.
2. Pat Bailey, "Grape Microbes Add to Wine's Distinctive Terroir," *University of California*, March 4, 2014, https://www.universityofcalifornia.edu/news/grape-microbes-add-wines-distinctive-terroir.

3. Ruth Williams and Full Profile, "Local Microbes Give Wine Character," *Scientist*, accessed August 8, 2022, https://www.the-scientist.com/news-opinion/local-microbes-give-wine-character-34788.

4. Jared M. Diamond, *Guns, Germs, and Steel: The Fates of Human Societies* (New York, NY: W. W. Norton & Company, Inc., 2017).

Notes to Music

1. Noam Sagiv and Jamie Ward, "Chapter 15 Crossmodal Interactions: Lessons from Synesthesia," *Progress in Brain Research*, 2006, 259–271, https://doi.org/10.1016/s0079-6123(06)55015-0.

2. Piesse G. W. Septimus, *The Art of Perfumery, and the Method of Obtaining the Odors of Plants* (London: Forgotten Books, 2017).

3. Lynne Peeples, "Making Scents of Sounds: Noises May Alter How We Perceive Odors," *Scientific American*, February 23, 2010, https://www.scientificamerican.com/article/making-scents-of-sounds-n/.

4. *Department of Experimental Psychology*, "Crossmodal Perception," accessed July 7, 2022, https://www.psy.ox.ac.uk/research/crossmodal-research-laboratory.

Notes to Minerality

1. Michael G. Tordoff, "Calcium: Taste, Intake, and Appetite," *Physiological Reviews* 81, no. 4 (January 2001): 1567–1597, https://doi.org/10.1152/physrev.2001.81.4.1567.

2. Alex Russan, "The Science of Salinity in Wine," *SevenFifty Daily*, February 9, 2022, https://daily.sevenfifty.com/the-science-of-salinity-in-wine/.

3. Russan, 2022.

4. Russan, 2022.

5. *WillyWeather*, "North Fork Wind Forecast," accessed November 30, 2022, https://wind.willyweather.com/ny/suffolk-county/north-fork.html.

6. Chris Macias, "When Smoke Gets in Your Wine," *University of California, Davis*, March 31, 2021, https://www.ucdavis.edu/climate/news/wine-climate-taint-solutions.

Notes to American Viticultural Areas (AVAs)

1. "TTB: Wine: American Viticultural Area (AVA)," TTBGov—American Viticultural Area (AVA), accessed April 10, 2023, https://www.ttb.gov/wine/american-viticultural-area-ava.

Notes to Native Yeast

1. Maria Tufariello et al., "Influence of Non-*Saccharomyces* on Wine Chemistry: A Focus on Aroma-Related Compounds," *Molecules* 26, no. 3 (US National Library of Medicine, January 26, 2021), https://www.ncbi.nlm.nih.gov/pmc/articles/PMC7865429/.

2. Nicholas A. Bokulich et al., "Associations among Wine Grape Microbiome, Metabolome, and Fermentation Behavior Suggest Microbial Contribution to Regional Wine Characteristics," *mBio* 7, no. 3 (US National Library of Medicine), accessed December 1, 2022, https://pubmed.ncbi.nlm.nih.gov/27302757/.

Notes to Wine and Biodynamics

1. Isabelle Legeron and Caroline West, *Natural Wine: An Introduction to Organic and Biodynamic Wines Made Naturally* (London: CICO, 2020).

2. Aaron Ayscough, "Not Drinking Poison," *Substack*, accessed 2022, https://notdrinkingpoison.substack.com/about.

3. Linda Chalker-Scott, "The Science behind Biodynamic Preparations: A Literature Review," *HortTechnology* 23, no. 6 (American Society for Horticultural Science, December 1, 2013), https://journals.ashs.org/horttech/view/journals/horttech/23/6/article-p814.xml.

4. Chalker-Scott, 2013.

5. Biodynamic Association, "Biodynamic Principles and Practices," Biodynamic Principles and Practices | Biodynamic Association, accessed December 7, 2022, https://www.biodynamics.com/biodynamic-principles-and-practices.

6. Lauren Johnson-Wünscher, "Opinion: Reconciling the Racism of Rudolf Steiner," *TRINK Magazine*, June 23, 2021, https://trinkmag.com/articles/opinion-reconciling-the-racism-of-rudolf-steiner.

7. Beach Combing, "Biodynamics and Nazi Market Gardens," *Beachcombing's Bizarre History Blog*, January 19, 2014, http://www.strangehistory.net/2010/11/15/biodynamics-and-nazi-gardens/.

8. "Barack Obama Quotes," BrainyQuote (Xplore), accessed December 5, 2022, https://www.brainyquote.com/quotes/barack_obama_168698.

Notes to Organic Winegrowing

1. "Global Organic Wine Market Growth Report, 2022–2030," *Grand View Research*, accessed November 30, 2022, https://www.grandviewresearch.com/industry-analysis/organic-wine-market-report.

2. Rachel Arthur, "The Organic Wine World Is in Full Expansion Mode—and Shows No Signs of Stopping!," beveragedaily.com (William Reed Ltd, February 8, 2019), https://www.beveragedaily.com/Article/2019/02/08/Organic-wine-market-continues-to-grow.

3. "4,500 Years of Crop Protection," *CropLife International*, April 4, 2017, https://croplife.org/news/4500-years-of-crop-protection/.

4. Albert Howard, *An Agricultural Testament* (London: Oxford University Press, 1972).

5. Miles McEvoy, "Organic 101: Organic Wine," USDA, February 21, 2017, https://www.usda.gov/media/blog/2013/01/08/organic-101-organic-wine.

6. "The National List of Allowed and Prohibited Substances," Agricultural Marketing Service, accessed November 30, 2022, https://www.ams.usda.gov/rules-regulations/national-list-allowed-and-prohibited-substances.

7. H. Eijsackers et al., "The Implications of Copper Fungicide Usage in Vineyards for Earthworm Activity and Resulting Sustainable Soil Quality," *Ecotoxicology and Environmental Safety* 62, issue 1 (Academic Press, May 10, 2005), https://www.sciencedirect.com/science/article/abs/pii/S014765130500031X.

8. Larry Perrine, "Copper Fungicide Effects on Soil Biota," interview by Richard Olsen-Harbich, 2022.

9. "Organic Wine Journal," Organic Wine Journal, March 21, 2022, http://www.organicwinejournal.com/.

Notes to Sustainability

1. World Commission on Environment and Development, *Our Common Future* (Oxford, England: Oxford University Press, 1987).

2. "What Is Sustainable Agronomy?" Sustainability | American Society of Agronomy, accessed December 2, 2022, https://www.agronomy.org/about-agronomy/sustainability.

Notes to Climate Change

1. Tony Greicius, "Study Projects a Surge in Coastal Flooding, Starting in the 2030s," NASA (NASA, July 7, 2021), https://www.nasa.gov/feature/jpl/study-projects-a-surge-in-coastal-flooding-starting-in-2030s.

2. K. A. Reed et al., "Forecasted Attribution of the Human Influence on Hurricane Florence," *Science Advances* 6, no. 1 (March 2020), https://doi.org/10.1126/sciadv.aaw9253.

3. Anthony Del Genio, "Earth's Changing Climate: How Will It Affect Viticulture?" interview by Richard Olsen-Harbich, 2022.

4. Ibid.

5. Ibid.

6. *Cornell Cooperative Extension*, "Growing Degree Days (GDD)," accessed September 2, 2022, https://ccesuffolk.org/agriculture/growing-degree-days-gdd.

7. Ibid.

8. Anthony Del Genio, "Earth's Changing Climate: How Will It Affect Viticulture?" interview by Richard Olsen-Harbich, 2022.

9. Richard Olsen-Harbich, "A Change Is Gonna Come," *Edible East End*, August 2, 2016, https://www.edibleeastend.com/2016/08/02/change-gonna-come/.

10. Suzanne Goldenberg, "Climate Change Will Threaten Wine Production, Study Shows," *Our World*, accessed December 9, 2022, https://ourworld.unu.edu/en/climate-change-will-threaten-wine-production-study-shows.

11. Ben Cook, "NASA Study Finds Climate Change Shifting Wine Grape Harvests in France and Switzerland," NASA, accessed October 8, 2022, https://www.giss.nasa.gov/research/news/20160321/.

12. Ben Cook, 2016.

13. S. Kaan Kurtural, "Global Warming and Wine Quality: Are We Close to the Tipping Point?" September 2021, https://www.researchgate.net/profile/S-Kurtural/publication/354906338.

14. Alice Wise interviews by Richard Olsen-Harbich, 2015–2022.

15. Larry Perrine interview by Richard Olsen-Harbich, 2015.

16. Ben Sisson interview by Richard Olsen-Harbich, 2005.

17. Robert Pincus, "Wine, Place, and Identity in a Changing Climate," *Gastronomica*, May 12, 2003, https://gastronomica.org/tag/terroir/page/2/.

18. Robert Pincus, 2003.

19. *Cornell Cooperative Extension*, "Precipitation Reports 2020–2029," accessed December 6, 2022, https://ccesuffolk.org/agriculture/growing-degree-days-gdd/gdd-from-previous-years/precipitation-reports-2020-2029.

Notes to No Women, No Wine

1. Jancis Robinson and Nick Lander, "The Feminisation of Wine," JancisRobinson.com, December 13, 2014, https://www.jancisrobinson.com/articles/the-feminisation-of-wine.

2. The Feminalise World Wine Competition / Concours mondial des vins féminalise, accessed 2022, https://www.feminalise.com/concours/presentation.php?lang=uk-en.

3. Women of the Vine & Spirits, August 29, 2022, https://www.womenofthevine.com/cpages/home.

4. Julie Brosterman, "Women & Wine," *Women & Wine*, accessed April 16, 2022, https://womenandwine.blogs.com/women__wine/.

5. Andrew Catchpole, "Massive Disconnect between Women and Wine World's Overwhelmingly Male Gatekeepers," *Harpers Wine & Spirit Trade News*, accessed 2021, https://harpers.co.uk/news/fullstory.php/aid/19447.

6. Julia Moskin, "The Wine World's Most Elite Circle Has a Sexual Harassment Problem," *New York Times* (October 29, 2020), https://www.nytimes.com/2020/10/29/dining/drinks/court-of-master-sommeliers-sexual-harassment-wine.html.

7. Court of Master Sommeliers, accessed January 6, 2023, https://www.mastersommeliers.org/about.

Notes to The Local Revolution

1. Tyler Colman and Pablo Päster, "Red, White, and 'Green': The Cost of Greenhouse Gas Emissions in the Global Wine Trade," *Journal of Wine Research* 20, no. 1 (2009): 15–26.

2. "Land Preservation," *Town of Southold, New York*, accessed December 30, 2022, http://southoldtownny.gov/115/Land-Preservation.

Notes to The French Connection

1. Larry Fuller-Perrine, *Maritime Climate Winegrowing: Bringing Bordeaux to Long Island: Proceedings of a Two-Day Symposium on the Production of Bordeaux Red Wine* (Geneva, NY: Communications Services, NYS Agricultural Experiment Station, 1988).

2. Larry Fuller-Perrine, 1988.

Note to Wines of Mass Vinification

1. Tim Bugher, "Wine and Juice Treating Materials and Processes for Domestic Wine Production," *TTB Alcohol and Tobacco Tax and Trade Bureau*, accessed 2022, https://www.ttb.gov/wine/treating-materials.

Notes to Our Sea-Washed, Sunset Gates

1. Immigration and Nationality Act of 1965 (Hart-Celler Act), *Immigration History*, September 27, 2019, https://immigrationhistory.org/item/hart-celler-act/.

2. "Farming on Long Island," *Long Island Farm Bureau*, accessed June 14, 2022, https://www.lifb.com/farming-on-li.

Notes to The Wild Oenophiles

1. *Town of Southold New York*, "Deer Management Program," accessed September 15, 2022, https://www.southoldtownny.gov/438/Deer-Management.

2. Greg Blass, "Wild Turkeys: Wonder of Nature or Annoying Pest?" *Riverhead LOCAL*, May 3, 2020, https://riverheadlocal.com/2020/05/03/wild-turkeys-wonder-of-nature-or-annoying-pest/.

Notes to Wine Ratings

1. Gary Rivlin, "Wine Ratings Might Not Pass the Sobriety Test," *New York Times*, August 13, 2006, https://www.nytimes.com/2006/08/13/business/yourmoney/wine-ratings-might-not-pass-the-sobriety-test.html.

2. David Darlington, "The Chemistry of a 90+ Wine," *New York Times*, August 7, 2005, https://www.nytimes.com/2005/08/07/magazine/the-chemistry-of-a-90-wine.html.

3. Iva Marinova, "Word of Mouth Marketing Statistics, Fun Facts & Tips in 2022," *Review42*, August 19, 2022, https://review42.com/resources/word-of-mouth-marketing-statistics/.

4. Iva Marinova, 2022.

Note to Myth-Busting

1. *Wikipedia*, "Cork (Material)," accessed March 20, 2023, https://en.wikipedia.org/wiki/Cork_(material).

Notes to Wine Futures

1. *Town of Southold New York*, "Land Preservation," accessed December 30, 2022, http://southoldtownny.gov/115/Land-Preservation.

2. *Town of Riverhead New York*, "Agricultural Lands," accessed December 30, 2022, https://ecode360.com/29708190.

3. Jacopo Mazzeo, "EU Grants Member States the Right to Use Resistant Hybrid Varieties in Appellation Wines," *Decanter*, December 14, 2021, https://www.

decanter.com/wine-news/eu-grants-member-states-the-right-to-use-resistant-hybrid-varieties-in-appellation-wines-470864/.

4. *ATTRA*, "Grapes: Organic Production," accessed November 26, 2022, https://attra.ncat.org/publication/grapes-organic-production/.

Notes to the Appendix

1. Federal Register, *Code of Federal Regulations*, accessed December 15, 2022, https://www.ecfr.gov/.

2. Ibid.

3. Ibid.

4. Ibid.

Bibliography

"Ages of Exploration." Accessed August 30, 2021, https://exploration.marinersmuseum. org/subject/henry-hudson/.

American Society of Agronomy. "Sustainability: What Is Sustainable Agronomy?" Accessed 2022, https://www.agronomy.org/about-agronomy/sustainability.

Amerine, M. A., and A. J. Winkler. "Composition and Quality of Musts and Wines of California Grapes." *Hilgardia* 15, no. 6 (1944): 493–675, https://hilgardia. ucanr.edu/Abstract/?a=hilg.v15n06p493.

Arthur, Rachel. "The Organic Wine World Is in Full Expansion Mode—and Shows No Signs of Stopping!" *beveragedaily.com*, February 8, 2019. https://www. beveragedaily.com/Article/2019/02/08/Organic-wine-market-continues-to-grow.

Ayscough, Aaron. "Not Drinking Poison." *Substack*, accessed 2022. https://not drinkingpoison.substack.com/about.

AZ Quotes. "Mark Twain Quote." Accessed December 1, 2022, https://www. azquotes.com/quote/358589.

Bailey, Pat. "Grape Microbes Add to Wine's Distinctive Terroir." University of California, March 4, 2014, https://www.universityofcalifornia.edu/news/grape-microbes-add-wines-distinctive-terroir.

BDA Certification. "Documentation for Demeter—Biodynamic." May 27, 2022, http://bdcertification.org.uk/index.php/documentation-biodynamic-demeter/.

Beachcombing's Bizarre History Blog. "Biodynamics and Nazi Market Gardens." January 19, 2014, http://www.strangehistory.net/2010/11/15/biodynamics-and-nazi-gardens/.

Becca. "The Effects of Climate Change on the California Wine Industry—My Latest Contribution to SOMM Journal." *Academic Wino*, August 10, 2016, http://www. academicwino.com/2015/10/climate-change-california-wine-somm-journal. html/.

Becker, Carl L. "Everyman His Own Historian (presidential address)." *American Historical Association*, accessed 2022, https://www.historians.org/about-aha-and-membership/aha-history-and-archives/presidential-addresses/carl-l-becker.

Beltrami, Edward J., Philip F. Palmedo, and Jose Moreno-Lacalle. *The Wines of Long Island*. Mattituck, NY: Amereon House, 2019.

Best Places. "Climate in Westchester County, New York." Accessed 2022, https://www.bestplaces.net/climate/county/new_york/westchester.

Biodynamic Association. "Biodynamic Principles and Practices." Accessed 2022. https://www.biodynamics.com/biodynamic-principles-and-practices.

Bland, Alastair. "An Upside to Climate Change? Better French Wine." *NPR*, March 21, 2016, https://www.npr.org/sections/thesalt/2016/03/21/470872883/an-upside-to-climate-change-better-french-wine.

Blass, Greg. "Wild Turkeys: Wonder of Nature or Annoying Pest?" *Riverhead LOCAL*, May 3, 2020, https://riverheadlocal.com/2020/05/03/wild-turkeys-wonder-of-nature-or-annoying-pest/.

Bokulich, Nicholas A., Thomas S. Collins, Chad Masarweh, Greg Allen, Hildegarde Heymann, Susan E. Ebeler, and David A. Mills. "Associations among Wine Grape Microbiome, Metabolome, and Fermentation Behavior Suggest Microbial Contribution to Regional Wine Characteristics." *mBio* 7, no. 3 (June 14, 2016), accessed December 1, 2022, https://pubmed.ncbi.nlm.nih.gov/27302757/.

BrainyQuote. "Barack Obama Quotes." Accessed December 5, 2022, https://www.brainyquote.com/quotes/barack_obama_168698.

Brosterman, Julie. Women & Wine (blog), accessed 2022, https://womenandwine.blogs.com/women__wine/.

Bugher, Tim. "American Viticultural Areas (AVAs)." *TTBGov*, accessed November 30, 2021, https://www.ttb.gov/wine/american-viticultural-area-ava.

Bugher, Tim. "Wine and Juice Treating Materials and Processes for Domestic Wine Production." *TTBGov*, accessed 2022, https://www.ttb.gov/wine/treating-materials.

Calheiros e Meneses, J. L. "The Cork Industry in Portugal by President, Junta Nacional Da Cortia, Portugal." The cork industry in Portugal, accessed 2022, https://www.researchgate.net/publication/233126030_Junta_nacional_da_cortica_1936–1972.

Carpenter-Boggs, L., A. C. Kennedy, and J. P. Reganold. "Organic and Biodynamic Management Effects on Soil Biology." *Soil Science Society of America Journal* 64, no. 5 (2000): 1651–59. https://doi.org/10.2136/sssaj2000.6451651x.

Catchpole, Andrew. "Massive Disconnect between Women and Wine World's Overwhelmingly Male Gatekeepers, Says Monika Elling." *Harpers Wine & Spirit Trade News*, accessed 2021, https://harpers.co.uk/news/fullstory.php/aid/19447.

Chalker-Scott, Linda. "The Science behind Biodynamic Preparations: A Literature Review." *HortTechnology* 23, issue 6 (December 2013): 814–15, https://journals.ashs.org/horttech/view/journals/horttech/23/6/article-p814.xml.

CLIMOD 2, accessed 2022, http://climod2.nrcc.cornell.edu/.

Coelho, Fábio C., Rosanna Squitti, Mariacarla Ventriglia, Giselle Cerchiaro, João P. Daher, Jaídson G. Rocha, Mauro C. A. Rongioletti, and Anna-Camilla Moonen. "Agricultural Use of Copper and Its Link to Alzheimer's Disease." *Biomolecules* 10, no. 6 (June 12, 2020): 897, https://www.ncbi.nlm.nih.gov/pmc/articles/PMC7356523/.

Colman, Tyler. "Red, White, and 'Green': The Cost of Greenhouse Gas Emissions in the Global Wine Trade." *Journal of Wine Research* 20, issue 1 (June 18, 2009): 15–26, accessed September 23, 2022, https://www.tandfonline.com/doi/full/10.1080/09571260902978493.

Cook, Ben. "NASA Study Finds Climate Change Shifting Wine Grape Harvests in France and Switzerland." *NASA: Goddard Institute for Space Studies*, accessed October 8, 2022, https://www.giss.nasa.gov/research/news/20160321/.

Cooperative Extension Association of Suffolk County. "Suffolk County Farm and Home Bureau Agricultural News, Volume 55." Riverhead, NY 1971.-

Cornell Cooperative Extension. "Growing Degree Days (GDD)." Accessed 2022, https://ccesuffolk.org/agriculture/growing-degree-days-gdd.

Cornell Cooperative Extension. "Precipitation Reports 2020–2029." Accessed 2022, https://ccesuffolk.org/agriculture/growing-degree-days-gdd/gdd-from-previous-years/precipitation-reports-2020-2029.

Cornell University. "Cornell Food Scientists Are Helping to Uncork the Mystery of 'Brett' Aroma in Wines." *ScienceDaily*, March 18, 1998, https://www.sciencedaily.com/releases/1998/03/980318075314.htm.

Court of Master Sommeliers (website), accessed January 6, 2023, https://www.mastersommeliers.org/about.

CropLife International. "4,500 Years of Crop Protection." April 4, 2017, https://croplife.org/news/4500-years-of-crop-protection/.

CU Boulder Today. "Toward More Sustainable Wine: Scientists Can Now Track Sulfur from Grapes to Streams." May 31, 2022, https://www.colorado.edu/today/2022/05/24/toward-more-sustainable-wine-scientists-can-now-track-sulfur-grapes-streams.

Current Results. "Days of Sunshine Per Year in New York." Accessed 2022, https://www.currentresults.com/Weather/New-York/annual-days-of-sunshine.php.

Darlington, David. "The Chemistry of a 90+ Wine." *New York Times Magazine*, August 7, 2005, https://www.nytimes.com/2005/08/07/magazine/the-chemistry-of-a-90-wine.html.

Data.gov. "Soil Survey Geographic (SSURGO) Database for Suffolk County, New York." November 29, 2020, https://catalog.data.gov/dataset/soil-survey-geographic-ssurgo-database-for-suffolk-county-new-york.

Department of Health: New York State. "The Suffolk County Comprehensive Water Resources Management Plan Team, Suffolk County Department of

Health Services." Accessed 2022, https://www.health.ny.gov/prevention/public_health_works/honor_roll/2016/suffolk_cwrmpt.htm.

Diamond, Jared M. *Guns, Germs, and Steel: The Fates of Human Societies.* New York, NY: W. W. Norton & Company, 2017.

Dozier, Crystal A., Doyong Kim, and David H. Russell. "Chemical Residue Evidence in Leon Plain Pottery from the Toyah Phase (1300–1650 CE) in the American Southern Plains." *Journal of Archaeological Science: Reports* 32 (2020): 102450, https://doi.org/10.1016/j.jasrep.2020.102450.

Dufour, Rex. "Grapes: Organic Production." *ATTRA*, accessed 2022, https://attra.ncat.org/publication/grapes-organic-production/.

Dunning, Josh. "Biodynamic Farming: Myths, Quacks and Pseudoscience." *Word on the Grapevine*, January 9, 2022, https://wordonthegrapevine.co.uk/biodynamic-viticulture-pseudoscience/.

Eijsackers, H., P. Beneke, M. Maboeta, J. P. E. Louw, and A. J. Reinecke. "The Implications of Copper Fungicide Usage in Vineyards for Earthworm Activity and Resulting Sustainable Soil Quality." *Ecotoxicology and Environmental Safety* 62, issue 1 (September 2005): 99–111, https://www.sciencedirect.com/science/article/abs/pii/S014765130500031X.

Environmental Protection Agency. "Sun Safety Monthly Average UV Index." Accessed 2022, https://www.epa.gov/sunsafety/sun-safety-monthly-average-uv-index#tab-9.

Federal Register Code of Regulations. "The Federal Register." Accessed 2022, https://www.ecfr.gov/.

Federal Register Code of Regulations. "The Federal Register: Title 27." Accessed 2022, https://www.ecfr.gov/current/title-27.

Feminalise World Wine Competition/Concours mondial des féminalise website, accessed 2022, https://www.feminalise.com/concours/presentation.php?lang=uk-en.

Figiel, Richard. *Circle of Vines.* Albany, NY: State University of New York Press, 2014, accessed 2022, https://sunypress.edu/Books/C/Circle-of-Vines.

Fishermap. "Long Island Sound Nautical Chart." Accessed 2022, https://usa.fishermap.org/depth-map/long-island-sound-ny-ct/.

Fuller-Perrine, Larry. *Maritime Climate Winegrowing: Bringing Bordeaux to Long Island: Proceedings of a Two-Day Symposium on the Production of Bordeaux Red Wine.* Geneva, NY: Communications Services, NYS Agricultural Experiment Station, 1988.

Garvies Point Museum and Preserve. "Geology of Long Island." Accessed 2022, https://www.garviespointmuseum.com/geology.php.

Goldberg, Howard G. "First 1973, and Then '88. Now, This Year." *New York Times*, July 6, 2008, https://www.nytimes.com/2008/07/06/nyregion/nyregionspecial2/06vinesli.html.

Goldenberg, Suzanne. "Climate Change Will Threaten Wine Production, Study Shows." *Our World*, accessed 2022. https://ourworld.unu.edu/en/climate-change-will-threaten-wine-production-study-shows.

Grand View Research. "Organic Wine Market Size, Share and Trends Analysis Report by Type (Red Organic Wine, White Organic Wine), by Distribution Channel (On-trade, Off-trade), by Region, and Segment Forecasts, 2022–2030." Accessed 2022, https://www.grandviewresearch.com/industry-analysis/organic-wine-market-report.

Greicius, Tony. "Study Projects a Surge in Coastal Flooding, Starting in 2030s." *NASA*, July 7, 2021, https://www.nasa.gov/feature/jpl/study-projects-a-surge-in-coastal-flooding-starting-in-2030s.

Hargrave, Louisa. "Center for Wine, Food and Culture." Center for Wine Food and Culture. Accessed 2022. https://you.stonybrook.edu/winecenter/symposiums/symposium2008/.

Howard, Albert. *An Agricultural Testament*. London: Oxford University Press, 1972.

Immigration History. "Immigration and Nationality Act of 1965 (Hart-Celler Act)." September 27, 2019, https://immigrationhistory.org/item/hart-celler-act/.

Jacobson, Aileen. "A Summer Place, Magnetic Year-Round." *New York Times*, May 4, 2012, https://www.nytimes.com/2012/05/06/realestate/cutchogue-li-living-in-a-summer-place-magnetic-year-round.html.

Johnson-Wünscher, Lauren. "Opinion: Reconciling the Racism of Rudolf Steiner." *TRINK Magazine*, June 23, 2021, https://trinkmag.com/articles/opinion-reconciling-the-racism-of-rudolf-steiner.

Krasilnikov, P. V., Juan-Jose Ibanez Marti, Richard Arnold, and Serghei Shoba. *A Handbook of Soil Terminology, Correlation and Classification*. London, England: Routledge, 2016.

Kurtural, S. Kaan. "Global Warming and Wine Quality: Are We Close to the Tipping Point?" September 2021, https://www.researchgate.net/profile/S-Kurtural/publication/354906338_Global_warming_and_wine_quality_Are_we_close_to_the_tipping_point/links/615381d339b8157d9003febf/Global-warming-and-wine-quality-Are-we-close-to-the-tipping-point.pdf.

Lallemand Oenology. "Copper Sulfate Use in Organic Vineyards: Are Wine Yeast and Bacteria Affected by This Fungicide?" Winemak-in. Accessed 2022. https://www.lallemandwine.com/en/north-america/expertise-document/the-new-winemaking-update-organic-version-copper-sulfate-use-in-organic-vineyards-are-wine-yeast-and-bacteria-affected-by-this-fungicide/

Legeron, Isabelle, and Caroline West. *Natural Wine: An Introduction to Organic and Biodynamic Wines Made Naturally*. London, England: CICO, 2020.

Long Island Farm Bureau. "Farming on Long Island." Accessed 2022, https://www.lifb.com/farming-on-li.

Long Island Farm Bureau. "Long Island Farm Bureau." accessed 2022. https://www.lifb.com/.

Macias, Chris. "When Smoke Gets in Your Wine." *UC Davis*, October 20, 2021, https://www.ucdavis.edu/climate/news/wine-climate-taint-solutions.

Marinova, Iva. "Word of Mouth Marketing Statistics, Fun Facts & Tips in 2022." *Review42*, August 19, 2022, https://review42.com/resources/word-of-mouth-marketing-statistics/.

Massa, Krysten. "Is Cutchogue Really the Sunniest Spot in New York?" *Suffolk Times*, June 4, 2016, https://suffolktimes.timesreview.com/2016/06/is-cutchogue-really-the-sunniest-spot-in-new-york/.

Mazzeo, Jacopo. "EU Grants Member States the Right to Use Resistant Hybrid Varieties in Appellation Wines." *Decanter*, December 14, 2021, https://www.decanter.com/wine-news/eu-grants-member-states-the-right-to-use-resistant-hybrid-varieties-in-appellation-wines-470864/.

McEvoy, Miles. "Organic 101: Organic Wine." *USDA*, February 21, 2017, https://www.usda.gov/media/blog/2013/01/08/organic-101-organic-wine.

Monell Chemical Senses Center, accessed May 24, 2022, https://monell.org/.

Moskin, Julia. "The Wine World's Most Elite Circle Has a Sexual Harassment Problem." *New York Times*, October 29, 2020, https://www.nytimes.com/2020/10/29/dining/drinks/court-of-master-sommeliers-sexual-harassment-wine.html.

NASA. "NASA Study Finds Climate Change Shifting Wine Grape Harvests in France and Switzerland." Accessed 2022, https://www.giss.nasa.gov/research/news/20160321/.

Nassau Suffolk Water Commissioners' Association (NSWCA). "Our Long Island Aquifers: The Basics." Accessed 2022, https://www.nswcawater.org/water_facts/our-long-island-aquifers-the-basics/.

National Centers for Environmental Information (NCEI). "Daily Summaries Station Details NY." Accessed 2022, https://www.ncdc.noaa.gov/cdo-web/datasets/GHCND/stations/GHCND:USW00094728/detail.

New Netherland Institute. "A Tour of New Netherland—Long Island." Accessed November 23, 2022, https://www.newnetherlandinstitute.org/history-and-heritage/digital-exhibitions/a-tour-of-new-netherland/long-island/.

Newsday. "Long Island: Our Story / Chapters 1–5." February 27, 2019, https://www.newsday.com/long-island/long-island-our-story-w53584.

Organic Wine Journal. Accessed March 21, 2022, http://www.organicwinejournal.com/.

Pacific Northwest Pest Management Handbooks. "Materials Allowed for Organic Disease Management." Accessed August 15, 2022, https://pnwhandbooks.org/plantdisease/pesticide-articles/materials-allowed-organic-disease-management.

Peeples, Lynne. "Making Scents of Sounds: Noises May Alter How We Perceive Odors." *Scientific American*. February 23, 2010, https://www.scientificamerican.com/article/making-scents-of-sounds-n/.

Piesse, G. W. Septimus. *The Art of Perfumery*. London, England: Forgotten Books, 2017.

Pincus, Robert. "Wine, Place, and Identity in a Changing Climate." *Gastronomica*, May 12, 2003, https://gastronomica.org/tag/terroir/page/2/.

Prince, William Robert, and William Prince. *A Treatise on the Vine: Embracing Its History from the Earliest Ages to the Present Day, with Descriptions of above Two Hundred Foreign and Eighty American Varieties: Together with a Complete Dissertation on the Establishment, Culture, and Management of Vineyards*. New York: T. & J. Swords, 1830.

Prince, William. "A Treatise on the Vine; Embracing Its History from the Earliest Ages to the Present Day, with Descriptions of above Two Hundred Foreign and Eighty American Varieties; Together with a Complete Dissertation on the Establishment, Culture, and Management of Vineyards . . ." *Biodiversity Heritage Library*. New York: T. & J. Swords [etc.], 1830, accessed 2022, https://www.biodiversitylibrary.org/bibliography/30498.

Puckette, Madeline. "The Bottom Line on Sulfites in Wine." *Wine Folly*, accessed 2022, https://winefolly.com/deep-dive/sulfites-in-wine/.

Puckette, Madeline. "Looking for Good Wine? Start with the Appelation." *Wine Folly*, accessed January 5, 2022, https://winefolly.com/deep-dive/looking-for-good-wine-start-with-the-appellation/.

Reed, K. A., A. M. Stansfield, M. F. Wehner, and C. M. Zarzycki. "Forecasted Attribution of the Human Influence on Hurricane Florence." *Science Advances* 6, no. 1 (2020). https://doi.org/10.1126/sciadv.aaw9253.

Reeve, Jennifer R., Lynne Carpenter-Boggs, John P. Reganold, Alan L. York, and William F. Brinton. "Influence of Biodynamic Preparations on Compost Development and Resultant Compost Extracts on Wheat Seedling Growth." *Bioresource Technology* 101, issue 14 (July 2010): 5658–5666, https://www.sciencedirect.com/science/article/abs/pii/S096085241000266X?via%3Dihub.

Reganold, John P. "Soil Quality and Profitability of Biodynamic and Conventional Farming Systems: A Review." *American Journal of Alternative Agriculture* 10, issue 1 (October 30, 2009): 36–45. https://www.cambridge.org/core/journals/american-journal-of-alternative-agriculture/article/abs/soil-quality-and-profitability-of-biodynamic-and-conventional-farming-systems-a-review/47CCCA63F2570B2D9EA6341BB05CF489.

Richter, Daniel K. *Facing East from Indian Country: A Native History of Early America*. Cambridge, MA: Harvard University Press, 2003.

Rivlin, Gary. "Wine Ratings Might Not Pass the Sobriety Test." *New York Times*, August 13, 2006, https://www.nytimes.com/2006/08/13/business/yourmoney/wine-ratings-might-not-pass-the-sobriety-test.html.

Robinson, Jancis, and Nick Lander. "The Feminisation of Wine." *JancisRobinson.com*, August 16, 2019, https://www.jancisrobinson.com/articles/the-feminisation-of-wine.

Russan, Alex. "The Science of Salinity in Wine." *SevenFifty Daily*, February 9, 2022, https://daily.sevenfifty.com/the-science-of-salinity-in-wine/.

Sagiv, Noam, and Jamie Ward. "Chapter 15 Crossmodal Interactions: Lessons from Synesthesia." *Progress in Brain Research* 155, part B (2006): 259–71, https://doi.org/10.1016/s0079-6123(06)55015-0.

Schmitt, Patrick. "Copper to Be the Biggest Issue for Wine Growers This Decade." *Drinks Business*, January 18, 2019, https://www.thedrinksbusiness.com/2019/01/copper-to-be-the-biggest-issue-for-vintners-this-decade/.

Science Daily. "Global Warming Pushes Wines into Uncharted Terroir." March 21, 2016, https://www.sciencedaily.com/releases/2016/03/160321123549.htm.

Sea Temperature Info. "Long Island Sound Water Temp." Accessed 2022, https://seatemperature.info/long-island-sound-water-temperature.html.

Sea Temperature Info. "Peconic Bay Water Temp." Accessed August 30, 2022, https://seatemperature.info/peconic-bay-water-temperature.html.

Sea Temperature Info. "Westhampton Water Temp." Accessed August 30, 2022, https://seatemperature.info/westhampton-water-temperature.html.

Sirkin, Leslie A. *Eastern Long Island Geology: History, Processes, and Field Trips.* Watch Hill, RI: Book and Tackle Shop, 1995.

Smith, Stuart. "Introduction." Biodynamics Is a Hoax website. Accessed 2022. https://biodynamicshoax.wordpress.com/.

Spence, Charles. "Crossmodal Perception." *Department of Experimental Psychology*, accessed 2022, https://www.psy.ox.ac.uk/research/crossmodal-research-laboratory.

Spooner, Alden. "The Cultivation of American Grape Vines, and Making of Wine." *WorldCat*, accessed 2022, https://www.worldcat.org/title/cultivation-of-american-grape-vines-and-making-of-wine/oclc/3803557.

Steen-Adams, Michelle. "Russell, E. 2001. *War and Nature: Fighting Humans and Insects with Chemicals from World War I to Silent Spring.* Cambridge University Press, Cambridge, UK and New York, New York, USA." Conservation Ecology 6, no. 2 (August 12, 2002), https://www.ecologyandsociety.org/vol6/iss2/art1/.

Suffolk County Government. "Departments: Water Resources." Accessed 2022, https://www.suffolkcountyny.gov/Departments/Health-Services/Environmental-Quality/Water-Resources.

Suffolk County Government. "Soils." Accessed 2022, https://www.suffolkcountyny.gov/Departments/Soil-and-Water-Conservation-District/Soils.

Tordoff, Michael G. "Calcium: Taste, Intake, and Appetite." *Physiological Reviews* 81, no. 4 (2001): 1567–97. https://doi.org/10.1152/physrev.2001.81.4.1567.

Town of Riverhead, New York. "Agricultural Lands." Accessed 2022, https://ecode360.com/29708190.

Town of Riverhead, New York. "Farmland Preservation Committee." Accessed December 30, 2022, https://www.townofriverheadny.gov/pview.aspx?id=3793&catid=0.

Town of Southold, New York. "Deer Management Program." Accessed September 15, 2022, https://www.southoldtownny.gov/438/Deer-Management.

Town of Southold, New York. "Land Preservation." accessed 2022, http://southold-townny.gov/115/Land-Preservation.

Tufariello, Maria, Mariagiovanna Fragasso, Joana Pico, Annarita Panighel, Simone Diego Castellarin, Riccardo Flamini, and Francesco Grieco. "Influence of Non-*Saccharomyces* on Wine Chemistry: A Focus on Aroma-Related Compounds." *Molecules* 26, no. 3 (January 26, 2021): 644, https://www.ncbi.nlm.nih.gov/pmc/articles/PMC7865429/.

United States Department of Agriculture (USDA). "The National List of Allowed and Prohibited Substances." Accessed 2022, https://www.ams.usda.gov/rules-regulations/national-list-allowed-and-prohibited-substances.

United States Department of Agriculture (USDA). "New York Online Soil Survey Manuscripts." New York Online Soil Survey Manuscripts | NRCS Soils. Accessed October 30, 2021, https://www.nrcs.usda.gov/conservation-basics/natural-resource-concerns/soil/soil-surveys-by-state.

van der Donck, Adriaen. *A Description of New Netherland.* Edited by Charles T. Gehring and William A. Starna. Translated by Diederik Willem Goedhuys. Lincoln, NE: University of Nebraska Press, 2010, accessed 2022, https://www.jstor.org/stable/j.ctt1dgn3z2.7.

Vanity Fair. "An American Original." October 6, 2010, https://www.vanityfair.com/news/2010/11/moynihan-letters-201011.

Verrazzano. "The Discovery of New York." accessed 2022, https://www.verrazzano.com/en/la-scoperta-di-new-york/.

Weather Spark. "Compare the Climate and Weather in Bordeaux, Ithaca, Tours, Riverhead, and Nantes." accessed August 30, 2022, https://weatherspark.com/compare/y/43632~22130~44094~25444~41180/Comparison-of-the-Average-Weather-in-Bordeaux-Ithaca-Tours-Riverhead-and-Nantes.

Wikipedia. "Long Island Sound." February 24, 2022, https://en.wikipedia.org/wiki/Long_Island_Sound.

Wilcox, Christie. "Mythbusting 101: Organic Farming > Conventional Agriculture." *Scientific American*, July 18, 2011, https://blogs.scientificamerican.com/science-sushi/httpblogsscientificamericancomscience-sushi20110718mythbusting-101-organic-farming-conventional-agriculture/.

Williams, Amber. "The Future's Not Straightforward. Neither Is Modeling It." *Scienceline*, December 15, 2011, https://scienceline.org/2011/12/the-future%E2%80%99s-not-straightforward-neither-is-modeling-it/.

Williams, Ruth. "Local Microbes Give Wine Character." *Scientist Magazine*, accessed 2022, https://www.the-scientist.com/news-opinion/local-microbes-give-wine-character-34788.

Willy Weather. "North Fork Wind Forecast." Accessed 2022, https://wind.willy-weather.com/ny/suffolk-county/north-fork.html.

Winkler, A. J., James A. Cook, William Mark Kliewer, and Lloyd A. Lider. *General Viticulture.* Edited by Laura Cerruti. Berkeley: University of California Press, 1974.

Women of the Vine and Spirits website. Accessed August 29, 2022, https://www.womenofthevine.com/cpages/home.

World Commission on Environment and Development. *Our Common Future.* Oxford, England: Oxford University Press, 1987.

Zubryd, Sascha. "Global Warming Could Significantly Alter the U.S. Premium Wine Industry within 30 Years, Say Stanford Scientists." Stanford University, July 13, 2011. https://www.losaltosonline.com/food/stanford-scientists-global-warming-could-alter-u-s-premium-wine-industry-within-30-years/article_eb282ee6-18ae-57aa-9e8f-32484ee3ec97.html.

Sources

Adams, Leon D. *The Wines of America,* 4th ed. New York, NY: McGraw-Hill, 1990.

Denton, Daniel. A *Brief Description of New York (1670).* New York, NY: Gowans, 1845.

Diamond, Jared M. *Guns, Germs, and Steel: The Fates of Human Societies* New York, NY: W. W. Norton & Company, 2017.

Huggett, Jennifer M. "Geology and Wine: A Review." *Proceedings of the Geologists' Association* 117, issue 2 (2006): 239–47.

Isachsen, Yngvar W., Ed Landing, Judy M. Lauber, Lawrence V. Rickard, and William B. Rogers. *The Geology of New York.* Albany, NY: New York State Museum, 1991.

Keller, M. *The Science of Grapevines: Anatomy and Physiology.* San Diego, CA: Academic Press, 2010.

Loubat, Alphonse. *The American Vine-Dresser's Guide.* New York: Carvill, 1827.

Palmedo, Philip, and Edward Beltrami. *The Wines of Long Island: Birth of a Region.* Richmond: Waterline, 1993.

Moreno-Lacalle, José. *The Wines of Long Island. Revised and Updated.* New Paltz, NY: Rivers Run By Press, 2019.

Prince, William R. *A Treatise on the Vine.* New York, NY: Swords, 1830.

Septimus, G. W. Piesse. *The Art of Perfumery, and the Method of Obtaining the Odors of Plants, Etc.* London, England: Forgotten Books, 2017.

Smart, R., and M. Robinson. *Sunlight into Wine: A Handbook for Winegrape Canopy Management.* Adelaide, SA: Winetitles, 1991.

Verrazzano, Giovanni da, *Journal of Expedition.* 1524, New York: Cosimo Classics, 2010.

Winkler, A. J., J. A. Cook, W. M. Kliewer, and L. A. Linder. *General Viticulture,* 2nd ed. Berkeley, CA: University of California Press. 1974.

Wood, Silas. *A Sketch of the First Settlement of the Several Towns on Long Island.* Brooklyn: Spooner, 1824.

Index

Catawba grapes, 183

cellar work, 42–43; careful handling, 82

Central American immigrant population, 146–147

Central Park, New York City, 27

Chardonnay, 77, 80, 82, 83, 152–154; harvest, 42

Chasselas, 179

Château Margaux, 75, 137, 157

Chateau Pichon Longueville Comtesse de Lalande, 137

Chenin Blanc, 179

Chile and Chilean wines: climate change, 121

cisgenesis, 182

cisgenic plants, 182–183

Cistercian monks, terroir and, 55

clarification: filtration, 93–94, 106; fining, 106

classic styles of wine: comparison to fads, 152–155

classification systems, wines, 55

clay, 15, 16, 19–20, 106

climate and weather, 26–28, 33–46, 47–49, 156, 199–205; acceptance, 112; comparison of terms "climate" and "weather," 33; definitions, 33; microclimate, 33–34; sea effects, 22, 35; South Fork, 27, 51–53; Winkler scale, 47; winter, 43–46, 156–157. *See also* frost-free days; Growing Degree Days (GDD); precipitation; sun and sunlight; temperatures; wind and wind events

climate change, 34, 117–125; air pollution and, 120; California, 119–120; classification of regions, 48; future trends, 181; Growing Degree Days and, 119–120; growing season and, 124; harvest and, 42; lower-alcohol wine and, 80; myth-busting, 171; precipitation and, 118–123, 181; South Fork/Hamptons, 52,

119–120; temperature and, 122–124, 181; terroir and, 123, 124

"Climate Change Decouples Drought from Early Wine Grape Harvests in France" (Cook), 121

clones and clonal/rootstock combinations, 78, 83, 153; future trends, 177, 182–183

collaboration among winemakers, 144; symposium on Old World/New World comparisons, 136–140. *See also* Long Island Sustainable Winegrowing (LISW) program; training and learning

Colman, Tyler, 133

Columbia University, 118, 121

Columbian Exchange, 13, 96, 186. See also *Phylloxera*

Columbian immigrant population, 146

Community Preservation Priority Plan, 178

competitions, wine, 129, 137

compost and composting, 96, 102–103, 135. *See also* natural wine movement and biodynamics

Connecticut: comparison to, 34, 35, 36, 43, 48; connections with, 50

Connecticut River, 14, 23

Conservation International, 121

Cook, Ben, 121, 125

"Cool New World," 2–3, 18, 74

copper and copper-based fungicides: organic winegrowing and, 96–98; sustainability and, 102

Corchaug people, 9, 12, 54, 192

Corey Creek Vineyard, 5

cork: natural cork, 90, 169; when tasting wine, 175

Cornell Agricultural Research Station, Riverhead, 70

Cornell Cooperative Extension, 130, 188; Suffolk County Grape Research Program, 122, 130

history and background of winemaking in North Fork region of Long Island, 2, 7–13, 191–194
Honduran immigrant population, 146–147
Housatonic River, 14, 23
Howard, Albert, 96
Howard Roto-Press, 74
Hudson, Henry, 7–8
Hudson Valley region, New York, 152
Hurricane Gloria, 25, 45
hurricanes, 25, 45; climate change and, 123
hybrids, 182
hydrogen sulfide, 83

ideologies. *See* natural wine movement and biodynamics; organic winegrowing and agriculture; sustainability and sustainable wine growing
immigrant population, winemaking and, 146–147
Immigration and Nationality Act of 1965, 146
inaugural luncheon, Obama's, 152
India, organic farming and, 96
indigenous grapes. *See* native grapes
indigenous peoples. *See* Native Americans
indigenous yeast, 74, 85–88
industrial wineries, lack of terroir, 141–142
insects: climate change and, 123; future trends, 181; spotted lantern fly, 181. See also *Phylloxera*
integrated pest management (IPM), 102
intervention philosophies, 76–77. *See also* additives; natural wine movement and biodynamics
iron, taste of, 62

irrigation, 35, 39; clay and, 20; climate change and, 122, 124; farm-to-table movement and, 134; West Coast vineyards, 134
Isabella (grapes), 183
isinglass: as fining agent, 106
Italy and Italian wines: ancient techniques, 95; climate change, 121, 125; Denominazione di Origine Controllata e Garantita (DOCG), 55; Homeric Greeks and Romans, 90; women in winemaking, 129. *See also specific regions*

Journal of Wine Research, 133

Kalish, Connie, 38
Kline Institute for Psychiatric Research, 59

labor: immigrant population, winemaking and, 146–147; shortages, 187
Labrusca grapes, 11; *Vitis labrusca,* 9, 11, 81, 212
Lagrien, 179
land and topography of Long Island, 14–17
land classes, 198–199
land use and development: evolution on Long Island, 160; farm-to-table movement, 133–134, 149; farmland preservation programs, 133–134, 150, 178; future trends, 177–178; nitrates in groundwater, 30
land values: Hamptons, comparison with, 51
Latino immigrant population, 146–147
lavender, effect on grapes, 65
learning. *See* training and learning
"legs": when tasting wine, 175
Leuconostoc bacteria, 83

Linnaean Botanic Garden, 10
LISW. *See* Long Island Sustainable
  Winegrowing (LISW) program
Lloyd aquifer, 29
loam topsoil, 19–21, 34, 51, 71, 197,
  198
local foods movement. *See* farm-to-
  table movement
Loire Valley comparisons, 19, 35, 178,
  179, 180
Long Island AVA, 215–217
Long Island Grape Growers
  Association, 69
Long Island Horticultural Research
  and Extension Center (LIHREC),
  42, 69, 70, 153, 199
Long Island Sound, 22–25, *25*;
  creation of, 16; salinity of, 23
Long Island Sustainable Winegrowing
  (LISW) program, 101–104, 144,
  187; Bordeaux symposiums, 138;
  farm-to-market movement and, 135
Long Island Wine Council, 70
Loubat, Alphonse, 10, 11
Loureira Blanca, 179
lower-alcohol wine: future trends,
  176–177. *See also* alcohol content
lye-ash, 90
Lynne, Michael, 101

magazines and wine industry
  publishing: wine ratings and,
  164–166; women in, 129. *See also*
  specific titles
Magothy aquifer, 29, 30
Malbec, 78, 83, 154, 180; Bordeaux
  comparisons, 138–139; soil
  influence, 20
malic acid, 64
Malo-lactic fermentation, 105, 106
Manhattan: connections with, 50;
  geography of, 15

manipulations. *See* intervention
  philosophies
maritime climate effect, 34, 37, 41, 43
"Maritime Climate Winegrowing:
  Bringing Bordeau to Long Island"
  (symposium), 137
marketing: social media and word of
  mouth, 165; women and, 129, 131
mass-produced wines, 141–142, 154
meditation, 111
Melon de Bourgogne, 78, 179
Mencia, 179
Merlot, 57, 77, 78, 83, 152–154,
  179, 180; Bordeaux comparisons,
  136–137, 138; climate change and,
  171; soil influence, 20
metals, taste of, 62
Mexican immigrant population, 146
microbial terroir, 55, 86
microclimate, 33–34
microflora: terroir and, 55, 86, 87
Middle Ages, women in winemaking,
  129
mildew, 12, 35, 97–98, 102, 182, 183
Mills, David, 55
mindfulness and winemaking, 111–
  113, 175
mineral oils, 98
minerality, 61–67; aroma and, 61–62;
  saline minerality, 62–64
Monell Chemical Senses Center, 62
Montauk Point, 15, 214, 215–216
Montaukett Nation, 9
Moraima, 62–63
moraines: Roanoke Point Moraine, 15,
  19, 23, 51; Ronkonkoma Moraine,
  15, 51. *See also* glaciers
Morrin, Tim, 27
Moynihan, Daniel Patrick, 48
Mudd, David, 13, 36, 44
Mudd, Stephen, 13, 81, 138, 139, 176
Mudd's Vineyard, 13, 24, 136–137, 176

North Fork, 24; climate, 199–205; climate change, 52; comparison to North Fork, 27, 50–53, 195–198, 203–204, 210–211; early vineyards, 9; frost-free days, 52–53; Growing Degree Days (GDD), 52; growing season, 25, 52; land classes, 198–199; origin of name, 190; precipitation, 51; red wines and, 52; soils of, 24, 51, 196–198; sunlight and, 27; temperatures, 24, 51, 53; weather, 24, 27, 51–53

Southold, New York, 9, 178, 191, 192, 195, 213, 214, 216

Spain and Spanish wines, 179; Denominación de Origen Protegida (DOP), 55. *See also specific regions*

spitting, delicate art of, 173, 175

Spooner, Alden, 9

spotted lantern fly, 181

spring: weather, 35, 36–38; winemaking, 36, 37

stainless steel tanks, 80

Stanford University, 121

starlings (birds), 161

Steiner, Rudolf, 91–92, 96

Stony Brook University, 118

style. *See* balance and style

subsoils, 20, 31, 34, 35, 124

succinic acid, 64

Suffolk County Agricultural News, 1971, 13

Suffolk County Department of Health Services (DHS): Groundwater Investigations Unit, 30–31

sugar: as additive, 76, 79; climate change and, 123

sulfur and sulfites, 83, 89–90, 93, 105, 107, 108–110; allergic reactions to, 109–110, 169; European wines compared, 97, 108–109; myth-busting, 169–170; organic

winegrowing and, 96–99; use in ancient Sumer, 95

Sumer, women and winemaking, 128

summer: weather, 38–40; winemaking, 39–40

sun and sunlight, 35; average sunny days, New York State, 28; North Fork region, 26–28; photosynthesis, 26, 39, 210; ultraviolet light penetration, 123; ultraviolet radiation index, 27–28; "wet light," 27. *See also* Growing Degree Days (GDD)

sustainability and sustainable wine growing, 100–104; contamination of ground water, 30; definition, 100; early proponents, 96; future trends, 177, 181; regenerative agriculture, 102. *See also* Long Island Sustainable Winegrowing (LISW) program

sweet wines: when tasting, 174

swirling, when tasting, 174

Switzerland: climate change, 121

symposium on Old World/New World comparisons, 136–140

synesthesia, 58–59

Syrah, 179

Taibo, Roberto, 62–63

tannin: as additive, 79

tartaric acid, 9

taste and tasting, 172–175; dos and don'ts, 173–175; local wines, 133; minerality, 62. *See also* description of wine; wine ratings

Tax and Trade Bureau (TTB), 68

Taylor, Chip, 88

technology. *See* equipment and technology

temperatures, 36; climate change and, 122–124, 181; future trends, 181; Long Island Sound, 23, 24; Peconic

virtual advertising, 165
viticulture, instruction in, 73
Vitis labrusca, 9, 11, 81, 212
Vitis vinifera, 77, 81; climate change and, 118; early attempts at growing, 9, 11–13, 193–194; effect of climate and soil on, 210–212; effect of sea and water on, 25; freezing temperatures and, 44; future, 182, 183, 194–195; *Phylloxera* and, 11–12
volatile phenols, 65

Wading River, 190, 191, 213, 214, 216
water: mass-produced wines and, 142; wells and well-testing, 30–32. *See also* aquifer; irrigation; precipitation; sea and seaside regions
A Way to Practical Settlement (Schwarz), 92
weather. *See* climate and weather
weed control, 39, 181; cover crops, 102–103, 135
wells and well-testing, 30–32
West Coast vineyards: irrigation, 134; overproduction in, 154; smoke taint from wildfires, 65. *See also* California wines and winemaking
Westchester Region, New York: comparison to, 36, 43, 48
"wet light," 27
white-tailed deer, 160–161
white wines, 52, 77, 78, 81; future trends, 179, 180
Wickham, John, 12, 193
Wiemer, Hermann, 44
wild fermentation, 83, 143–144
wild turkeys, 161, 162
wildfires, smoke taint from, 65
wildlife, 160–162; birds, 41–42, 161, 162

wind and wind events, 35, 36; climate change and, 123; effect on salinity, 63–64; Hurricane Gloria, 25, 45; hurricanes, 25, 45, 123; maritime climate effect, 34, 37, 41, 43
Wine Advocate, 164
Wine and Spirits, 164
wine classification systems, 55
wine competitions, 129, 137
Wine Enthusiast (magazine), 166
Wine Market Council, 129
wine ratings, 152, 163–167; women in, 129–130
Wine Spectator, 80, 152, 164
winemakers and winemaking, 72–78; childrearing analogy, 90–91; craft brewing industry compared, 143–145; decision making, 82; immigrant population and, 146–147; innovation and, 144; parenting/winemaking comparison, 126–127; physical toil, 74–75; raw materials, 82; risk involved, 143–144; sommeliers compared, 73; terroir, appreciation of, 84; traditional practices, 75–76; training and learning, 72–73, 75, 84; vineyard management, 83, 138; women in winemaking, 128–131; yield, 82
Winkler, A. J., 47
Winkler scale, 47
winter: climate change and, 123–124; insects and, 123; weather, 43–46, 156–157
Wisconsin Glacier, 15
Wise, Alice, 122, 153
women in winemaking, 128–131
Women of the Vine Alliance, 130
Women & Wine, 130
word of mouth (WOM), 165